W9-DIJ-355

VOLT RUSH

VOLT RUSH

The Winners and Losers in
the Race to Go Green

HENRY SANDERSON

ONEWORLD

A Oneworld Book

First published by Oneworld Publications in 2022
Reprinted twice in 2022

Copyright © Henry Sanderson 2022

ISBN 978-0-86154-375-5
eISBN 978-0-86154-376-2

Typeset by Geethik Technologies
Printed and bound in Great Britain by Clays Ltd, Elcograf S.p.A.

Oneworld Publications
10 Bloomsbury Street
London WC1B 3SR
England

Stay up to date with the latest books,
special offers, and exclusive content from
Oneworld with our newsletter

Sign up on our website
oneworld-publications.com

MIX
Paper from
responsible sources
FSC® C018072

To Claudia and Jamie

Contents

Introduction

'We have to scale battery production to crazy levels that people can't even fathom today.'

Elon Musk, Tesla CEO[1]

In mid-2020 as the world settled into lockdown, we decided to get an electric car. Our son was seven months old and I had begun to think seriously about his future and whether the world would act quickly enough to mitigate the threats of climate change. Shares in Tesla were soaring, and the carmaker was on the cusp of becoming the most valuable in the world, despite the fact it made a fraction of the eleven million cars a year produced by Toyota. Tesla's Model 3 had been the UK's bestselling car that April and our friends had started to take the plunge. I knew that our son's future would depend on decisions we made now, not later. Even as the pandemic had halted global economic activity the news was still grim: a Siberian heatwave had pushed up global temperatures to their second-highest on record. The well-known impacts of climate change stemming from our consumption of fossil fuels had not taken a break. So, I placed our old two-door petrol car for sale and started to look for electrics to rent. Google informed me that one popular search topic was: 'Will petrol cars be worthless?'

Electric cars were the ethical consumer choice. It was a seductive idea: we could change the world by slightly altering our current lifestyle with a marginal amount of sacrifice. Buying a green investment fund instead of an index tracker; an electric car instead of a petrol one. The BBC informed me that the way we choose 'to travel to the office, or even to pop to the shops, is also one of the biggest day-to-day climate decisions we face'.[2] The car leasing company I used promised a world where cars 'go hand in hand with the environment'. I imagined myself charging my car in the future and wondered how we would start to perceive petrol cars. Would driving one start to seem like an act of wanton vandalism to the planet, a thoroughly anti-social provocation?

For seven years I had lived in Beijing at the tail-end of China's thirty-year car boom. I would still wake up with memories of what it had felt like to live there: the dry grating taste at the back of my throat, the hot breath behind my mask (long before Covid) and the tightness in the centre of my chest. I remember watching the red taillights on giant ring roads flowing into a dusk the colour of dirty sink water and feeling unable to escape the city of over twenty million. On some days Beijing felt like a postcard from the end of the planet. China's car demand seemed like an unstoppable juggernaut – an urge to consume that would eat up whole cities, and then require new cities to be built for the cars.

The growth of China's car market in my lifetime is alarming. To make a meaningful difference to climate change we will need to scale up the electric car fleet rapidly. The current stock of electric cars globally is around ten million, less than twice the number of cars in Beijing and only one percent of the global total. There are over one billion cars globally on the roads. We will also need to replace buses and trucks with electric versions, as well as ships, ferries and even planes. All this will require batteries on a scale unimaginable a few years ago. Tesla's South African-born founder Elon Musk had built a vast battery 'Gigafactory' in the desert of Nevada to supply his electric cars as well as the batteries to store renewable sources of energy. But he was not alone: across China a new factory was being built every week in 2020.

Through my work as a *Financial Times* journalist, I had covered the raw materials electric cars needed: the lithium, cobalt, nickel and copper, as well as aluminium and steel. The harder I looked at the supply chain and who was responsible for these metals, the more I understood the shift in economic power that this transition would bring. While Musk and Tesla took all the media attention and limelight, there was a shadow world of billionaires who were also set to get rich. A gold rush had begun.

In the Democratic Republic of the Congo I saw private jets landing at the small green and white airport in the mining town of Kolwezi while all around children and families mined for cobalt by hand; in Chile I stood in the searing heat of the Atacama Desert overlooking giant pools the size of Manhattan where lithium was extracted; and in China I visited battery factories and lithium plants running twenty-four hours a day using coal-fired power next to fields where buffalo roamed freely. All this was part of the electric car supply chain. A host of companies were even planning to mine the deep sea for these minerals, opening up one of the last unexplored wildernesses to extraction.

Every day, we rely on metals and minerals to power our iPhones and transmit our electricity. Digital technologies give us a sense that we live in an ethereal economy untethered to the material world. In fact, we are mining more minerals than at any time in our history, a dependence that is only set to increase.* Despite talk of artificial intelligence, the internet of things, and an imminent takeover by robots, our societies have in many ways not moved on from the practices of the past, when the need for oil drove Europeans to carve up the Middle East.

---

* Analysts at the American brokerage firm Bernstein calculated that for every one percent you want to grow GDP, you must increase mined volumes by two percent.

The consequences of the transition will not just be economic – they will also be environmental. Extracting and processing these minerals requires large amounts of energy and pollutes local ecosystems. This fact is often hidden in debates about a transition to renewable energy and electric cars. Every product we use has contributed to global emissions, both through extraction of raw materials and manufacturing. Mining is estimated to contribute around ten percent to global carbon emissions. It is an unavoidable industry: to make steel we need coal, and to make batteries we need lithium, the world's lightest metal with the highest electrochemical potential. The electric vehicle (EV) revolution is green at its core but there are many choices to be made in the way it is executed that will affect both the environment and global power dynamics.

Green energy evangelists tend to assume that a fossil-free future will be without conflicts. Bill McKibben, a well-known activist, wrote that 'if the world ran on sun, it wouldn't fight over oil'.[3] This idea was repeated by Tony Fadell, the creator of the iPod, who told *Wired* magazine: 'If we have energy storage technologies that are very cheap and very efficient, then we're going to see wars stop, because no one is going to be fighting over oil reserves anymore.'[4] Yet the demand for raw materials to build our clean energy infrastructure is as geopolitical as the age of oil. Countries that are able to become a part of these new clean energy supply chains will benefit, while those who cannot will suffer.

The transition to clean energy has already deepened geopolitical tensions between the West and China. Over the past decade, China has obtained a dominant position in both clean energy technologies such as batteries and solar cells and the raw material supply chains that underpin them. While many of these minerals are not rare in the earth's crust, they are mostly processed in China into usable forms. It's likely the lithium and cobalt in my car were mined in Australia and the Congo yet processed in

China and turned into a battery by a Chinese company.* The electricity was probably generated from solar panels produced by a Chinese company from polysilicon made using coal-fired power in Xinjiang province. And the copper was almost certainly refined in one of China's hundreds of copper smelters. Without China, we cannot move towards a green, battery-powered future. As we move towards greater energy independence, these supply chains remain our greatest vulnerability.

This book tells the story of these supply chains and the characters behind them. The story involves some of the world's most secretive natural resource companies; private Chinese companies buying up mines in Chile, Australia and Indonesia; and the largest car companies in the world such as Tesla and Volkswagen. My hope is that the book will equip readers to ask the right questions about our transition away from fossil fuels. The resources we need are buried in the earth's crust; so what environmental and social cost are we willing to bear to extract them? Without scrutiny, abuses will remain hidden behind a veil of corporate 'greenwash' from companies further down the supply chain. We are at the beginning of the electric car revolution, which means we have a unique opportunity as consumers to push companies to do the right thing and manage the downsides involved. We shouldn't be hostile to green technologies but we shouldn't be naive either. The oil age has left a long scar on the twentieth century. We should make sure that the industries of our green future do much better.

---

* In this book, references to 'the Congo' in the period from 1964 onwards always refer to the Democratic Republic of the Congo (Congo-Kinshasa) rather than Congo-Brazzaville.

1

The Battery Age

'The spice must flow. The new spice.'

Elon Musk, Tesla CEO[1]

In late September 2020 Elon Musk strode out onto a stage in the bright California sunlight to address a car park full of Teslas at the company's annual shareholder meeting in Fremont. Dubbed 'Battery Day', the event had been hyped up by Musk for months (he had promised it would be 'insane'). Due to Covid-19 attendees sat alone behind the steering wheels of bright shiny Model 3 and Model Y cars wearing masks. They greeted Musk with a ripple of discordant honks of their horns, as if they were frustrated commuters stuck in traffic. 'Well this is a new approach. We've got the Tesla drive-in movie theatre basically,' Musk said.[2] The attendees had every reason to be happy: Tesla's shares had soared during the pandemic, making it the most valuable car company in the world, even though it accounted for only one percent of the global car market. At its current valuation each Tesla car they sat in was worth around $1 million. It was a remarkable turnaround from the last few years, when Tesla had come close to bankruptcy as it struggled to increase production of its Model 3 electric vehicle. Tesla, which had started life as a fast-moving Silicon Valley start-up that made cars for the rich, had become a proper automotive manufacturer in one of the most cut-throat and low-margin businesses in the world. It had succeeded in breaking the logjam that had held electric vehicles up for over a century by creating a product that people actually wanted to buy. Along the way it had created a devoted following of fans who hung on Musk's every word and tweeted about their Tesla cars.

Battery Day marked the next step in Musk's life mission (which also included human colonisation of Mars). Strutting the stage in a black T-shirt with a microphone clenched in his hand, Musk outlined Tesla's plan to cut battery costs in half and produce a $25,000 mass market electric car. Honk! Honk! For all the hype and excitement, the electric car was still too expensive to compete with petrol. It was an ambition reminiscent of Henry Ford's launch of the Model T car over a hundred years earlier, which had ushered in the motoring age by making cars affordable for the working man. Cars at the time had been luxury items but Ford had been determined to get the price below the average yearly wage. Just as Ford had pioneered the moving assembly line to lower costs, Musk needed to scale up battery production. Tesla would need to increase battery production one-hundred-fold by 2030 – enough to make around twenty million cars a year, he said.

That was just for Tesla. By the time of Battery Day, almost every car company in the world – from General Motors to Volkswagen – had pledged to go electric. And more than twenty countries had also announced future bans on the sale of petrol and diesel vehicles.

Then Musk dropped a bombshell. Tesla had acquired the rights to a 10,000-acre plot in Nevada where it planned to extract lithium, a metal that was the key ingredient for every electric vehicle battery, using simple table salt. There was enough lithium in the US to electrify the entire US vehicle fleet, Musk said. 'We take a chunk of dirt out of the ground, remove the lithium and then put the chunk of dirt back where it was. It will look pretty much the same as before. It will not look like terrible. It will be nice,' he explained. 'There's so much damn lithium on earth it's crazy.'[3]

Yet to meet Tesla's targets would require four times the amount of lithium that the world currently produced. The International Energy Agency predicted that demand for lithium was set to grow thirty-fold by 2030 and more than a hundred times by 2050. The US produced almost no lithium, cobalt and nickel, and little of the copper needed for charging stations and the electrical grid. The same was true for Europe.

Musk's lithium venture was not likely to produce any metal for years. Instead, his remarks were designed to jolt the notoriously sleepy mining industry into action. In the audience were executives from the two largest lithium companies in the US. It was reminiscent of remarks a hundred years earlier by Ford's key lieutenant Charles Sorensen: 'If others would not provide enough steel for our needs, then we would. It was just as simple as that.'[4]

In the meantime, Musk would have to rely on the green barons – the companies that controlled the emerging clean energy supply chain. A hundred years ago Ford's Model T had created fortunes for the early oil drillers and refiners, leading to the creation of the global oil industry and some of the world's largest companies. Now Musk had kickstarted a similar raw material rush. 'The spice must flow … the new spice,' Musk said, referring to the 1965 science fiction novel *Dune*, which detailed the struggle for control of a planet that produced the spice necessary for space navigation and the extension of life. Now that spice was a small group of metals – lithium, cobalt, copper and nickel.

The battery age had begun.

2

Dashed Hopes:
The Troubled History
of the EV

'Electricity is the thing. There are no whirring and grinding gears with their numerous levers to confuse. There is not that almost terrifying uncertain throb and whirr of the powerful combustion engine. There is no water-circulating system to get out of order – no dangerous and evil-smelling gasoline and no noise.'

Thomas Edison[1]

'Within a year, I hope, we shall begin the manufacture of an electric automobile. I don't like to talk about things which are a year ahead, but I am willing to tell you something of my plans.

'The fact is that Mr. Edison and I have been working for some years on an electric automobile which would be cheap and practicable. Cars have been built for experimental purposes, and we are satisfied now that the way is clear to success. The problem so far has been to build a storage battery of light weight which would operate for long distances without recharging. Mr. Edison has been experimenting with such a battery for some time.'

Henry Ford[2]

I drove down to collect our electric car on a clear, cold December day, just after the UK's second lockdown. The roads were busy with the first taste of freedom – a brief opening leading up to Christmas. Car ownership provided a safe way to go out in the pandemic, removing the need to mix with other households. It struck me that the car had returned to its original promise – the ability to go wherever you wanted, when you wanted. It was a concept that had captured American culture and came to define the Western world. I had grown up with it. The electric car now promised all the benefit and none of the guilt that had latterly come to define car ownership. The shift in the UK's energy mix, and therefore the energy that would power my car, was remarkable. The share of coal in the UK electricity mix had fallen from forty percent in 2013 to two percent. On many days it was zero. We now burned less coal than in 1882, when the first coal-fired power station was built. My car would be charged by energy from the wind and the sun. It was an enticing idea – the thought of driving on a renewable resource.

Before I had got rid of my petrol car, I had taken it into a garage in a narrow street in east London run by Jak, an energetic and entrepreneurial owner. He opened up the bonnet to examine the engine. My cylinders were misfiring, he told me. I gazed at the engine as if it were an exotic animal that was about to go extinct. We were surrounded by the noise and familiar reek of engines. Jak had spent his career fixing engines and it was clear he loved them. 'Engines are really very simple when you understand them,' he said. I asked him about electric cars, and he sighed. 'Electrics are OK but they will have their own drawbacks,' he said. 'There are teething problems.'

I was about to find out. On my journey to collect my Tesla, I was accompanied by my father, whose life had been defined by oil. He was born in Kirkuk, in Iraq, where my grandfather worked for the Iraqi Petroleum Company, a consortium of four of the world's largest oil companies, for many years – staying there after the 1958 revolution by Iraqi officers that overturned the country's Western-backed monarch. It was an era when Western oil companies dominated the global oil markets and divvied up the Middle East amongst themselves. Such efforts had enabled the rapid growth of car culture in Europe and America. Like my father I had grown up in the era of the car – the growth in distance travelled and car ownership shot up in the second half of the twentieth century. It had been, as researchers from MIT noted in 2001, a 'golden age' for mobility.[3]

We arrived at Thorpe Park, a place I had visited frequently as a child to ride the log flume – but now the theme park was temporarily closed, and Tesla had set up stall instead. Row upon row of bright shiny Teslas that had arrived from California sat in the winter sun. We queued up outside an office to get our keys, which turned out to be two black key cards. The man behind the counter told me the car's software would cut out temporarily when we left the car park but we were not to worry as we could still drive. His face made clear that any further questions were unnecessary. He pointed to the car park and told me to head off. No one showed us to the car or told us how it worked. 'Thank you for accelerating the transition to sustainable energy,' a notice said. My father and I walked out into the cold, found our car and entered with some trepidation. I had to scan a QR code and grapple with the large touch screen, which immediately checked for a software update. I grabbed a young man who was walking by to get his help. I asked where the back windscreen wiper was and he laughed. 'Teslas don't have those,' he said.

Then off I went, silently, as if something was missing. It just started! I was gliding along the road. I was so used to the combustion engine's presence, its noise and throbbing, that at first I thought I had lost something profound. There was no engine to warm up the car or melt the frost. Instead, a woman's voice guided me through every turn at the roundabout as I scrolled through the large screen searching for a radio station. For a moment I felt as if I was in an upsized golf cart. But as I turned onto the motorway, I soon forgot those thoughts as the car accelerated frighteningly quickly. I realised the American inventor Edison had been right: electricity is the thing. In 1909 Edison had told a friend over lunch: 'this scheme of combustion in order to get power makes me sick to think of – it is so wasteful.'[4]

Because of the software included in a Tesla, many writers had started to mourn the passing of the car age. To the writer Matthew Crawford, cars, especially autonomous self-driving ones, would now be nothing more than computers on wheels. Silicon Valley executives 'seek to make everything idiot-proof, and pursue this by treating us like idiots'.[5] In the *New Yorker* the writer Bruce McCall, a former advertising executive for the car industry, sounded a similar elegy, writing that he had been lucky to have experienced the 'romance of driving at its fervent peak'.[6] 'Mine was close to the last generation of car nuts,' he wrote. 'Since my time working on car ads, automobiles have morphed into emotionally neutered large appliances, competing more on entertainment than performance, dulling risk with technological interventions that replace the need for judgment. This is good for safety and inarguably progressive – but it's heading into a tomorrow where we'll all be guests in our automated, self-driving blobs.'

But to me these writers missed a critical point: the importance of an electric car was not in the electronics, or the software, as fancy as these were. The exciting development was the battery – the thousands of cells lying at the bottom of the car, wrapped in an aluminium case. It was improvements in the battery that had made the electric car possible. It meant that I, and not a tech billionaire, could go electric and drive two hundred miles without worrying. It marked a hard-won victory over the internal combustion engine.

*

In the summer of 1896 Thomas Edison, the man responsible for the first workable lightbulb and the phonograph, was dubbed the 'Wizard of Menlo Park'. Inventions seemed to flow uninhibited from his brain on an almost daily basis. On a single day when he was just over forty, Edison had noted down a hundred and twelve ideas for possible devices, including a mechanical cotton picker and an electrical piano as well as 'Ink for the Blind'.

Henry Ford, in contrast, was an unknown engineer from Michigan who worked at the Edison Illuminating Company in Detroit, part of Edison's electricity empire, helping to maintain the steam engines that generated electricity. In his spare time, Ford worked on his first petrol car, the Quadricycle, in a workshop outside his home. The prototype, however, had not received a positive reception in Ford's hometown after he rolled it out one rainy morning in June 1896. The vehicle had a two-cylinder, four-horsepower engine under its bench seat and a tiller for steering. But as he drove his four-wheeled motorised car around the streets of Detroit people thought it was a nuisance, 'for it made a racket and scared horses. Also it blocked traffic.'[7]

Ford idolised Edison and, at a conference held at the Oriental Hotel in Manhattan Beach, Brooklyn, that summer, Ford managed to take some pictures of the grand man sleeping on the hotel's veranda. The meeting had brought together electrical technicians and executives from all the Edison electrical companies across the US. On 12 August, the third day of the conference, Ford got a lucky break when he managed to attend the conference banquet with some of the leading players in America's electricity industry. Edison was the honoured guest. The delegates discussed electric cars and storage batteries but then, during a lull in the conversation, Ford's boss from Detroit announced that 'this young fellow here has made a gas car.'[8] The mood in the room must have turned – it was a conference about electricity after all. Someone at the table asked Ford for more details, whereupon he began to explain his petrol-powered Quadricycle, which he had spent three years building. Edison, who had bad hearing, strained to listen, so Ford moved closer to him. Edison continued to prod Ford with questions, so Ford sketched out the details for him on the back of a menu card. After he had finished, Edison brought his fist down on the table with a loud bang and told him to keep up the good work. Edison's encouragement meant the world to the young Ford; the next day Edison invited him to take the same train back to the city with him. Ford returned to Detroit full of enthusiasm to get going on making a second model of his car.

Despite his encouraging words to Ford, however, Edison himself still believed in the superior potential of electric vehicles. He had previously built a three-wheeled electric car in 1895.[9] But he had found batteries wanting. Edison had called the existing rechargeable lead-acid battery a 'catch-penny, a sensation, a mechanism for swindling by stocking companies'.[10] By 1900, Edison could not resist the lure of inventing a better battery himself. So it was that in May of that year, at the beginning of an optimistic new century, Edison stood on a street on the West side of Manhattan, waiting for the ferry to Jersey City. The streets were bustling with horse-drawn traffic, and the air stank of urine and manure. For two hours Edison stood and jotted ideas in his notebook:

Limited loads. Congestion. Resulting decay and expense there-from…

Solution:- Electrically driven trucks, covering one-half the street area, having twice the speed, with two or three times the carrying capacity... Development necessary:- Running gear-easy. Motor driver-easy. Control-simple. Battery-(?)[11]

That question mark symbolised the start of what would be a nine-year quest by Edison to find a better battery. It would be one of the most demanding periods of his career, involving multiple setbacks. Edison believed that inventing a battery was just a matter of finding the right materials. Lead-acid batteries, which had been invented by Gaston Planté in France in 1859, were the dominant rechargeable technology of the time, but they were big and clunky, and leaked sulphuric acid from the electrolyte. Edison envisioned the same properties that battery scientists pursue today: a battery that could store large amounts of energy for its weight; that could withstand being charged and discharged many times; and that could be driven around in any conditions. 'If Nature had intended to use lead in batteries for powering vehicles,' Edison said, with his typical optimism, 'she would not have made it so heavy.'[12]

Edison launched his quest at the right time. At the turn of the century the automotive industry was still up for grabs. Adverts in the newspapers for automobiles were a mix of steam, petrol and electric vehicles, which all vied for consumers' interest. There was the Perry Louis Electric, the Columbia Electric Runabout and the Riker Electric Tricycle. It was an age of discovery and invention and it seemed natural that battery technology would also advance rapidly to meet the needs of the 'horseless carriage'. Petrol cars had no obvious advantage: they were loud and polluting and they had to be started by a heavy crank, which could break an arm if it kicked back. Steam cars, on the other hand, while powerful, had to be regularly refilled with water. Electric vehicles seemed the best bet, at least for inter-city travel: they were simpler and they did not require any gears.

In fact in 1900 petrol cars accounted for twenty-two percent of total vehicles in the US, electrics thirty-eight percent and steam vehicles forty-nine percent.[13] In public transportation electric streetcars had already started to replace horse-drawn ones following the installation in 1887 of an electric streetcar system in Richmond, Virginia. A number of companies in New York also offered electric taxi services. From 1897 to 1912 the Electric Vehicle Company provided electric vehicles for hire in and around New York City and became the largest vehicle manufacturer and the largest owner and operator of motor vehicles in the US.[14] The company envisioned a nationwide fleet of electric taxicabs and the provision of mobility as a service, a similar concept to Uber today.

Edison applied his prodigious work ethic to the rechargeable battery. He aimed for a battery that had triple the energy density of contemporary lead-acid ones. By the start of 1901, Edison had more than a hundred staff helping him with his new battery. By May the battery was officially unveiled to the world at the annual meeting of the American Institute of Electrical Engineers. After experiments with cadmium Edison had settled on nickel for the positive electrode and iron for the negative electrode, and a potassium hydroxide electrolyte. Edison also used tiny flakes of graphite – which decades later would be the key to the lithium-ion battery – in the positive electrode, mixed in with nickel hydrate. The press was ecstatic, deeming the news to be as game-changing as any of Edison's inventions. 'The latest achievement of Edison is probably destined to work as great changes in its way as did the electric light,' the *Rochester Democrat and Chronicle* said.[15] In 1903, the same year Ford founded his eponymous car company, Edison boasted to a reporter that he would build an electric car that would 'be able to beat, or at any rate keep up with, any gasoline machine on a long run'.[16]

The stage seemed set for a battery revolution. The *New York Times* said that the 'Edison battery seems destined to work a speedy revolution and to bring the oft-predicted and much-desired banishment of the horse from our city streets near to realization'.[17] With a range of almost a hundred miles, the new electric vehicle would be as suitable in the country as in the cities due to the increase in the number of power stations around the country, the paper predicted. Edison agreed: 'Yes the new battery will settle the horse – not at once but by degrees,' he said. 'The price of automobiles will be reduced.'[18]

But Edison and the press had got ahead of themselves. The truth was his battery was not quite ready despite the multitude of tests Edison had subjected it to. It leaked electrolyte through microscopic pores and the cells lost their capacity quickly. Customers quickly started to complain. Edison was forced to make the painful decision to recall all the batteries on the market at his own cost, shut down production, and start again.

Edison returned to the lab, where he managed two teams who worked twenty-four hours a day to solve the problems. It would take until 1909 before Edison had perfected the battery – the robust A-12 nickel storage battery that was encased in a steel case, with two protruding poles for the positive and negative electrode. 'At last the battery is finished,' Edison wrote in the summer of that year.[19] He had spent over a million dollars of his own money on the venture.

But the delay in getting the improved battery to market turned out to be fatal. In the years Edison spent perfecting his battery, the internal combustion engine had gone from strength to strength, raising the bar against which batteries would have to compete. A year earlier Ford had unveiled his Model T, which quickly became a mass-market car. Edison's grand announcements about his earlier battery had fed a feeling that electric vehicles constantly disappointed people's expectations. The 'unwarranted promise by the daily newspapers of a 200-mile battery has proved a serious obstacle to the introduction of electric vehicles', one electric vehicle enthusiast noted in 1909.[20] By the end of the decade the balance of power between Edison and Ford had changed. Ford, now forty-eight, was a multimillionaire car tycoon who was one of the richest men in America. Petrol cars had won the race.

*

For a brief period at the end of the nineteenth century, as the historian Kirsch wrote, the electric vehicle 'was the dominant technology'. The outcome of total dominance of petrol cars was far from preordained. At a minimum it would have made sense to have different technologies for different purposes – electric vehicles for inner city travel and petrol cars for longer distances. 'The prospect of a privately owned, mass-produced motor vehicle for every American adult man and woman would have seemed nothing less than lunacy,' he wrote.[21]

Kirsch's investigation into the history of electric vehicles, which was published in 2000, left him pessimistic, like many other writers at the time, about their future. At the time, General Motors had recalled its electric car, the EV1, which initially used lead-acid batteries just as cars had done in 1900. It was a recall that left many owners distraught and disappointed. To Kirsch it was just the latest chapter in the electric vehicles' capacity to disappoint expectations.

Just as Kirsch was indulging his pessimism, however, improvements in a new technology – the lithium-ion battery – were changing the equation for electric vehicles. In his book Kirsch briefly mentioned the industry but went on to dismiss it. 'It has been nearly a century since Edison first promised a better battery, but in many respects we are still waiting for it.'[22]

The year after the publication of Kirsch's book, I went to visit an electronics store on Tottenham Court Road. The shop's owner showed me a white rectangular box that could store hours of music. It seemed clunky and strange-looking to me. The owner said it was called the iPod, handing it to me to hold. The digital mobile age had arrived and buried within the devices, a critical enabler, was the lithium-ion battery. Invisible to the consumer, the reach and impact of these batteries would be profound. Among the beneficiaries would be the electric vehicle. Everything was about to change.

3

The Breakthrough: The Lithium-Ion Revolution

On 10 December 2019 at a white tie banquet in Stockholm, British chemistry professor Stanley Whittingham rose to the stage to accept the Nobel Prize in chemistry as part of an international team that helped invent the lithium-ion battery in the 1970s and 1980s. The Nobel Academy had finally recognised the achievement, saying that lithium batteries had 'laid the foundation of a wireless, fossil fuel-free society, and are of the greatest benefit to humankind'. In the audience was John Goodenough, then ninety-six, who pioneered the battery at Oxford University in the 1980s and was the oldest person ever to win the Nobel Prize. Whittingham quoted remarks by Thomas Edison, who more than a hundred years ago had also struggled to invent a life-changing battery. 'There is a way to do it better. Find it,' Edison had said. 'I would like to think that we found it and with a systematic scientific approach,' Whittingham added. Science did not recognise disciplinary boundaries, he added, mentioning his fellow winners Goodenough and Akira Yoshino, an engineer from Kyoto University who had worked for Asahi Kasei Corporation, which made a commercial lithium battery with Sony. 'Colleagues, we hope that our discovery will let all of us work together to build a cleaner environment, make our planet more sustainable, and help mitigate global warming, thereby leaving a cleaner legacy to our children and grandchildren,' Whittingham finished.[1]

It was a sentiment echoed by Goodenough, who, despite his age, was still working every day at the University of Texas, helping to discover new battery chemistries. Tall and bushy-eyebrowed Goodenough had told reporters: 'We need to get burning fossil fuels off the highways and freeways of the world and focus on global warming. If you're going to have renewable energy ... you need a battery where you can store it.'[2]

A few months before the Nobel ceremony Whittingham, who had been born in Nottingham during the Second World War, had visited Volkswagen in Germany to view the company's new battery pilot centre. 'Fifty years ago when we started out, with these cars behind me, it would [have been] science fiction,' a grey-haired Whittingham said in front of an ID.3 electric car, Volkswagen's first mass-market offering. 'But now we've made it. So I think we've got to believe that we can do it. We must believe in the unbelievable. Then we can make it happen.'[3]

Goodenough and Whittingham had waited a long time for batteries to fulfil their true potential to solve the planet's excessive reliance on oil, gas and coal. It was a hope that was crystallised in the 1970s, when the world feared that oil was running out. It was a sentiment that was inspired by an unlikely source – ExxonMobil.

*

The 1970s began in the US with the first Earth Day, the start of a nationwide movement of speeches and protests in which an estimated twenty million Americans participated, galvanising a new environmental movement. Air pollution in cities such as Los Angeles had become infamous, prompting city officials to take action. Industrial pollution was also reaching a tipping point across the country – Cleveland's Cuyahoga River caught fire in

1969 due to oil-soaked debris burning on the surface, and even in Washington there was a coal-burning plant a few blocks from the Capitol.[4] The US government listened. Eight months later the first Clean Air Act was passed and President Richard Nixon established the Environmental Protection Agency. Three years later the 1973 Arab oil embargo and the Iranian revolution led to widespread fears about the future of oil supplies. Global car companies from Toyota to Daimler-Benz were all working on electric cars. 'It was obvious already in 1970 that our dependence on foreign oil was making the country as vulnerable as the threat of ballistic missiles from Russia,' Goodenough wrote in his memoir.[5] In 1977, the founder of Earth Day, Denis Hayes, published a book that summed up the mood: *Rays of Hope: The Transition to a Post-Petroleum World*. At the same time, scientists were warning about the effects of global warming on the planet, which had coalesced into a broad international consensus by the end of the decade.

This backdrop spurred renewed interest in batteries. As we saw in the previous chapter, they were not a new technology, but their roots lie even further back than the days of Edison and Ford. Batteries are a way of storing energy via chemical means. Most batteries consist of three main parts: a cathode (the positive electrode), the anode (the negative electrode) and an electrolyte. The chemical reaction between the two electrodes is harnessed to provide electricity. Batteries were first invented at the end of the eighteenth century following a dispute between two Italian scientists over what made a dissected frog's legs twitch. Luigi Galvani, an Italian surgeon who worked at the Academy of Sciences in Bologna, believed that animals generated a unique electrical charge in their muscles. Galvani conducted many experiments on frogs' legs using a machine known as the Dollond that created an electric spark. One day, instead of touching a wire to the frog's leg, an assistant 'gently touched the point of a scalpel to the medial crural nerves of this frog. Immediately all the muscles of the limbs contracted.'[6] The experiment had shown that the legs contracted without any direct connection to the Dollond machine. Galvani continued to experiment and published a full account in 1791. Galvani still believed that he had generated 'animal electricity'. But Alessandro Volta, a professor of physics at the University of Pavia, read Galvani's reports and decided to replicate the experiment. He found that two different metals in contact with each other would make the frog's legs twitch. Volta concluded that it was the contact between the two different metals that generated the electricity. He assembled what would become known as a 'Voltaic pile', a stack of two different metals on a wooden base, separated by blotting paper and connected by a wire. A current of electrons flowed through the wire from the more reactive to the less reactive metal. The difference in chemical reactivity between the metals generated volts – the force of electricity (and a term named after Volta). For the first time in history, Volta had generated a continuous, stable supply of electricity. Batteries would remain the main source of electricity until the late nineteenth century, powering new inventions including the telegraph, which enabled long-distance communication for the first time.

Over a hundred years would pass before the next breakthrough in battery technology. In 1859, the Frenchman Gaston Planté developed the lead-acid battery, which, as mentioned earlier, enabled batteries to be recharged for the first time. Electrons flowed in the other direction inside the battery with the application of electricity. Lead-acid batteries dominated the market for decades and were later used in petrol-powered cars, helping to power the lighting and provide the spark to help start the engine.

In 1967 Joseph Kummer and Neill Weber at the Ford Motor Company invented a sodium-sulphur rechargeable battery that used a solid ceramic electrolyte, instead of the typical liquid electrolyte. The same year Ford had unveiled a test electric car, the Comuta, which had been built in England. Using lead-acid batteries, the car could only drive forty miles at twenty-five miles an hour. The new battery promised to change that, allowing a car to drive up to eighty-two miles on a single charge. 'We're convinced this is the real answer,' Jack Goldman, director of Ford's scientific laboratory, said.[7] Yet the battery would only operate when both electrodes were molten, at over 300°C, making it impracticable for commercial use in vehicles. Like many such promised break-throughs over the years, the project fell by the wayside.

Still, the announcement helped catalyse research into bat-teries and especially the movement of ions (charged atoms) in solids rather than liquids. A year after Ford's announcement, Whittingham, who had just finished his DPhil at Oxford, had begun to investigate such materials while holding a fellowship at Stanford University. Whittingham and others studied the conduc-tivity of beta-alumina, a commercially available material that Ford had used for its battery electrolyte. In 1971, Whittingham won a Young Author Award from the Electrochemical Society for his work on the material. On the back of that research, he was hired by ExxonMobil to work on alternative energy projects just before the 1973 oil crisis hit.

'I joined Exxon in 1972 and early the next year I got sent on a course where they were predicting oil production was going to peak before 1990,' Whittingham recalled. Exxon modelled its corporate research division in Linden, New Jersey, on Bell Labs, the AT&T-funded research laboratory. Exxon worked on a range of technologies from solar panels to nuclear reactors. 'If you go back BP did the same thing – the oil companies wanted to become energy companies,' Whittingham explained. He ini-tially researched superconductors, materials that could conduct electricity at room temperature, using a material called tantalum disulphide. Whittingham discovered that the material could also store energy. Yet the tantalum was too heavy, so he settled on a lighter metal: titanium. He then needed to find a suitable negative electrode to make a battery, and settled on lithium because it is the lightest alkali metal and gives the highest voltage.

Whittingham knew that lithium might work from pictures he had seen of Japanese fishermen, who used non-rechargeable lithium batteries at night to help them see their nets. Silvery-white and lustrous, lithium reacts violently with water and is so volatile that it can only be found in compounds in nature. But it gives up an electron easily, making it a perfect choice for batteries. His team managed to produce a rechargeable lithium battery that worked at room temperature. The Exxon battery cells used a titanium disulphide cathode (the positive electrode) and a lithium metal anode (the negative electrode) separated by a liquid electrolyte which contained a lithium salt in an organic solvent. In early 1973 Whittingham was sent to Exxon's headquarters in Manhattan to present his findings to a subcommittee of the board of directors. He had to hunt around for the right building 'because in those days Exxon didn't really put their name in big letters on the buildings', he recalled. 'I was told you've got five or ten minutes at most.'

Whittingham's pitch was successful and within a few days Exxon decided to enter the battery business. 'They said "you're going to invest in it and build a development team" and eventually they got into manufacturing,' Whittingham said. 'They looked at research then like drilling wells,' he said. 'It's high risk but a few percent will pay off.' The batteries were exhibited at the Chicago electric vehicle show in 1977, where they powered a motorcycle's headlamp, and the oil company opened a manufacturing facility in Branchburg in East Jersey. 'I think there was surprise: one, that there was a large lithium-ion battery and, two, that Exxon was doing it,' Whittingham told me. Exxon started producing small button-sized batteries to power a Swiss watch but the oil company had ambitions to move into electric vehicles and convert a VW car into an electric vehicle.

However, as oil prices fell in the following years, the appeal of electric vehicles declined. In 1980, the free marketer Ronald Reagan defeated Jimmy Carter to win the presidency. Exxon sold off Whittingham's technology, and research into lithium-ion batteries largely returned to the domain of academics. 'The oil crisis of the 1970s went away, top management had changed and they came in one day and looked at a lot of these things and said "you mean the market is not 100 million dollars a year. Why are we in it?" And I think they got out of most of the alternative energy businesses of that time,' Whittingham recalled. Between 1977 and 1986, Exxon also pioneered studies of carbon dioxide and climate change, before that programme too fell by the wayside. By the end of the 1980s the oil giant had taken a dramatic turn and 'worked at the forefront of climate denial'.[8]

The reality was that Whittingham's lithium battery was wholly unsuitable for a car: it had only a small voltage of less than 2.5 volts, and had been prone to catching fire. The lithium metal anode, although stable with the electrolyte, on charging formed dendrites, needle-like crystals that could penetrate through the separator, short-circuiting the cell and causing it to ignite.

It was John Goodenough, whom we met at the start of the chapter, a German-born American physicist working in Oxford in the early 1980s, who took on the challenge of finding a better battery material. He invented a new combination of cathode materials that improved the voltage and amount of energy a lithium-ion battery could store. Goodenough's team at Oxford and then the University of Texas came up with the three leading cathode chemistries that are still in use today. Without this scientific research, we would not be buying electric cars today. 'We were patiently doing basic science all those years, that is paying off twenty to thirty years later,' Arumugam Manthiram, a materials scientist who was born in a small village in India and later left to study with Goodenough in Oxford in 1985, told me. Manthiram, who is known as Ram, then followed Goodenough to Austin, where Goodenough still works in the office next door, even as he approaches his hundredth birthday.

Goodenough was born in Jena, Germany, in July 1922 and grew up amid meadows and woodlands in Woodbridge, Connecticut, seven miles north of Yale where his father worked in the history faculty. His parents' marriage was 'a disaster' and Goodenough grew up fearing that he would be abandoned. He also struggled to learn to read. He left home at the age of twelve to go to boarding school, and later attended the prestigious Groton School, where he 'responded positively to the structure and discipline' and made friends.[9] He was never homesick, he wrote in his autobiography, *Witness to Grace*. In his final year of the school, his father divorced his mother to marry his research assistant. But Goodenough studied diligently for his exams and made it to Yale, just as Europe fell deeper into war.

At Yale Goodenough discovered philosophy and science, reading *Science and the Modern World* by Alfred North Whitehead. He managed to graduate just after he was called up for active service in 1943. Entering the army as a meteorologist he spent time analysing weather patterns in the Azores and rose to the position of captain.

At the war's end Goodenough won a place to study physics at the University of Chicago, helped by his old Yale mathematics professor. While toying with the idea of becoming a Christian minister, he joined the MIT Lincoln laboratory after graduation, where he worked on random access memory for early computers. He also started to work with transition metal compounds, which would be key to lithium-ion batteries. In 1970, like Whittingham, he began to think more about renewable energy. But Goodenough was told that the MIT lab was an Air Force laboratory and his research needed to focus on the needs of the military. Discouraged, he decided to leave. He was offered a position to set up a research laboratory in Tehran and travelled to Iran in 1974. But on his way home a letter arrived from Oxford inviting him to apply for the position of professor and head of the inorganic chemistry laboratory there. Goodenough applied and got the position, even though he was not a traditionally trained chemist. He was now free to work on the problem of energy.

Goodenough was inspired to find a better cathode material by an undergraduate thesis in 1978 on the structure of metal oxides. He set his team to work finding suitable metals that would work in a battery without the use of a reactive lithium metal anode. The material Goodenough developed, lithium cobalt oxide, had good structural stability and could withstand the repeated insertion and extraction of lithium ions as the battery charged and discharged. It 'solved two major challenges associated with the sulphide cathodes pursued in the 1970s. It enabled not only a substantial increase in the operating voltage from <2.5V to ~4V but also the assembly of a cell without the need to employ a metallic lithium anode.'[10]

Despite its promise and the lingering impacts of the 1970s energy crisis, however, no battery company was interested in a patent on the technology. Oxford University was also not interested in patenting academic research – a far cry from today when the university jumps at such opportunities. Goodenough arranged for the Atomic Energy Research Establishment Laboratory in Harwell in the UK to file a patent on his behalf, but their lawyers made him sign his rights away.[11] He didn't realise there was 'any other option', he wrote in his autobiography.[12] The lab later received billions of pounds from the cathode patent, but Goodenough got nothing – not even a donation to his Oxford college.[13]

It was Japan that ended up reaping the benefits of the research conducted at Oxford. Japan's electronics industry had anticipated the rise of mobile products and was desperate for a lightweight rechargeable battery. It was in this environment that Akira Yoshino, who also won the Nobel Prize, delivered the final piece of the puzzle. Yoshino had joined Japan's Asahi Kasei Corporation in 1972 and began to look into the use of polyacetylene, a conducting polymer, as a battery material. He was looking for a suitable cathode to pair it with when he came across Goodenough's paper. It was just what he was looking for. But he soon found that polyacetylene would not make batteries small enough, so he decided to use carbon. After a long search, he found the right sort of carbon with a particular crystalline structure that could hold lithium ions – petroleum coke, a by-product of the oil industry. He also added a separator that provided a 'shutdown function in the case of abrupt heat generation'.[14] He completed his lithium-ion battery in 1985. It was commercialised in 1991 by Sony and a year later by A&T Battery, a joint venture between Asahi Kasei and Toshiba. It was just in time – as the *New York Times* had noted in 1989: 'the consumer's appetite for portable electricity has leaped ahead of the ability of batteries to store it.'[15]

Small and light, lithium-ion batteries revolutionised the consumer electronics market, allowing Sony and others to produce handheld video cameras, laptops and, later, phones and tablets. When the battery provided power, positively charged lithium ions moved from the negative anode through the electrolyte to the positive cathode, where they embedded in the cobalt-oxide metal invented by Goodenough. The electrons released by the reaction, however, moved through an external circuit, which powered the device. When the battery was charged, the opposite happened. Cobalt was critical to providing the stability of the cathode material, allowing it to withstand the repeated insertion and withdrawal of lithium ions.

In 2007, worldwide ownership of mobile phones passed the one billion mark; lithium batteries had cemented their position at the centre of our digital lives. 'It takes a long, long, time … many years, for any new technology to become commercial, and many more years to catch up on the cost curve,' Mark Newman, a battery analyst, told me. Yoshino put it perfectly: 'Basically, when a new innovation is born, you need two breakthroughs,' he said in early 2021. 'One is in basic research, the other is actually mass-producing it. A breakthrough isn't real until you have both.'[16] The performance of lithium-ion batteries has improved steadily over the decades, with the amount of energy stored in a litre-sized pack more than tripling to over 700 watt-hours per litre. Costs have also fallen remarkably, dropping by ninety-seven percent since Sony introduced the first lithium-ion battery, according to scientists at MIT.[17]

As well as powering the rise of the digital mobile age, the breakthrough in lithium-ion batteries was set to change the game in electric vehicles. Lower lithium-ion battery costs meant electric cars were on the cusp of being affordable and becoming mass-market items. Electric car batteries used the same formula that Goodenough had discovered, with the addition of metals such as nickel and manganese to the cathode to cut costs and boost the amount of energy the battery could store.

Key to that transformation was China, where generous government subsidies and Beijing's ambitions to be a leader in electric cars had created a battery superpower.

4
China's Battery King

'"The typhoon is coming, even the pigs can fly." But are the pigs flying? Once the typhoon passes, what is the situation for the pigs that are left? ... Have we ever thought that if foreign companies come back in the second half of the year, can we still sleep blindfolded? Will the state protect uncompetitive businesses?'

Robin Zeng, founder of CATL,
to employees in early 2017[1]

In late 2019 residents of the German town of Arnstadt, which has a population of 28,000, awoke to find diggers breaking ground on a new factory on the outskirts of town, on the site of a defunct solar panel plant. Stretching over twenty-three hectares (around 100 football fields), the $2 billion plant was Germany's first large-scale battery 'Gigafactory' with the capacity to pump out enough batteries for hundreds of thousands of electric cars every year. Germany invented the internal combustion engine in 1876 and makes the finest luxury cars in the world. The car industry helped power Germany's post-war *Wirtschaftswunder* with brands such as BMW, Mercedes-Benz and Audi becoming symbols of reliability and engineering expertise. But the Arnstadt factory was not being built by a German carmaker. Instead it was being financed and constructed by a little-known Chinese company founded only eight years earlier in the mountainous eastern fishing town of Ningde, which is better known for its tea plantations and yellow croaker. The company, Contemporary Amperex Technology, or CATL, had already struck deals to supply batteries to Volkswagen and BMW as they sought to reinvent themselves and move away from producing internal combustion engines. It also had an agreement to supply batteries to Daimler's electric buses and trucks. German carmakers were turning to China for the core technology needed to switch their entire industry to electric cars. In other words, China was providing them with the means for their own survival. Batteries, the most expensive part of an electric vehicle, were critical to its success as a mass-market item, and CATL had substantially lowered their costs. 'The rest of the world is playing catch-up,' Bob Galyen, an American from Indianapolis who had worked as chief technology officer at CATL in Ningde, told me.

By 2019 Germany's largest industry faced a stark transition. To meet European Union climate change targets, the country's carmakers needed to go electric or they faced the prospect of large fines from Brussels. So carmakers such as Volkswagen had begun to make bold promises about the number of electric cars they would produce and the tens of billions of euros they would spend. Yet European carmakers had no homegrown battery production or any presence in the broader battery supply chain. For years they had not really believed in electrification enough to invest money in large battery factories. That had left the ground open for Elon Musk's upstart Tesla Motors. As CATL's factory was breaking ground, Tesla was in negotiations to build a Gigafactory outside Berlin – another encroachment on Germany's home turf. The move was called a 'declaration of war' by the *Frankfurter Allgemeine Zeitung* newspaper.[2] German carmakers had no choice but to head to Asia with their chequebooks open to buy batteries in bulk and take stakes in Chinese battery companies. 'Our competitors are not based in Ulm or Munster,' Wolf-Dieter Lukas, state secretary in the German Federal Ministry of Education and Research, said. 'They are based in South Korea and China. The clear message is: we have to be at the forefront when the battery technology of the future is developed.'[3] It was a reversal of fortunes for Europe. Germany was used to being a supplier to China of advanced manufacturing; but now China had moved up the value chain and was a competitor. It was a trend that Europe had done little to stop: in 2016 Chinese home appliance maker Midea had bought the German robotics company Kuka, whose robots were critical to making batteries. CATL's founder Robin Zeng was fully aware of the historical significance of his company's deal to supply the Mercedes-Benz brand: 'Mercedes invented the car more than 130 years ago and has developed its technology with countless innovations. This combined with CATL's expertise in battery, will be a decisive step in both parties' electrification strategies,' he said.[4]

How had a Chinese company that few people had heard of managed to defeat the German carmakers at their own game? By 2020 CATL was supplying almost every electric carmaker, including Tesla, giving the company a dominant position in the transition away from fossil fuels. It also supplied batteries to some of Germany's competitors – a number of Chinese start-ups such as Nio and Xpeng. These companies, which were listed on the US stock exchange, had started to export cars to Europe. MG, the British car company owned by China's state-owned SAIC, already sold its ZS EV using CATL batteries in the UK. CATL also owned stakes in lithium projects in Argentina and Australia, a nickel project in Indonesia and a cobalt deposit in the Democratic Republic of the Congo – giving it access to the resources it needed. By consolidating its hold over the battery industry, China hoped to build a world-leading electric car industry. For China's ruling Communist Party electric cars helped solve two major problems: they reduced air pollution in the country's cities, and helped cut China's reliance on oil imports. The round-faced and diminutive Zeng was worth $51.6 billion and was the twenty-fifth richest person on earth, according to the 2021 Bloomberg Billionaires list.

The company had created more billionaires than either Google or Facebook, and was already worth more than Volkswagen. Germany had made a 'strategic error of neglecting the research and development of batteries', according to Stanford professor Fritz Prinz, who was born in Austria. 'Perhaps it was thought that batteries would only be needed for smartphones and other portables, which was a mistake.'[5]

The headquarters of CATL lie on the edge of Ningde, a stone's throw from ponds where farmers raise carp. The giant factory is next to a street of cheap noodle restaurants and vehicle repair shops frequented by migrant workers. Inside the factory, battery parts move silently on automated conveyor belts. There are few people – and none of the armies of migrant workers that typified the Chinese boomtown factories of the 2000s. Once an impoverished city with little but tea plantations and mountains, Ningde is well known in China due to President Xi Jinping's stint as Party secretary in the city from 1988 to 1990. At the time it was a demotion for Xi, who was moved from the busy coastal city of Xiamen to Ningde after his father, Xi Zhongxun, a high-ranking Party member, had refused to support a crackdown on the liberal reformer Hu Yaobang. A year later Hu's death was the spark for student protests in Tiananmen Square, which resulted in the violent crackdown by the People's Liberation Army in June 1989 that put an end to hopes for reform of China's political system. Xi was in Ningde as China was convulsed by the nationwide protests.

The same year, a young man called Robin Zeng made his way to southern China, to the bustling coastal city of Dongguan in southern Guangdong province near Hong Kong, which was embracing capitalism and openness to the world, despite the widespread crackdown on political speech that followed the Tiananmen violence. For an ambitious young man in 1989 moving to Dongguan was like heading to the centre of the world, a place becoming connected to global supply chains where workers lived in crowded dormitories and could watch Hong Kong TV. In the city, which a few years earlier had been farmland and rice paddies, foreign investors were being encouraged by the local government to invest in manufacturing enterprises. Dongguan attracted significant Taiwanese, Hong Kong and other foreign investment in factories, and acted as a magnet for Chinese migrant workers (Dongguan's population doubled in the 1980s). It was a frontier town: a world of factories and prostitutes where migrants came on one-way train tickets to make clothes or toys. In the evenings its streets were flooded with young workers finishing their shifts, full of hopes and dreams. Zeng found work at a Hong Kong company called SAE Magnetic that made magnetic recording heads for computer hard drives – an industry Dongguan would come to dominate. Thin-film magnetic heads allow computers to be smaller and store more data. The company had been recently bought by Japan's electronic giant TDK.

It was already a significant change from Zeng's childhood. He had been born in the small mountain village of Lankou outside Ningde during the chaos of the Cultural Revolution in 1968. Born into a farming family, Zeng was naturally intelligent and at the age of seventeen left school to study engineering at Shanghai Jiao Tong University, escaping the small town of Ningde. A top student, he received a doctorate in condensed matter physics from the Chinese Academy of Sciences in Beijing.[6] After graduating he joined a state-owned company in Fujian province, where he could have led a comfortable life with an 'iron rice bowl', as working at state-owned companies was known – a job that would have left his parents very satisfied and proud. But the entrepreneurial Zeng was not content to idle away his hours in the sleepy state sector and left after just three months.

Zeng stayed in Dongguan for ten years, rising to become the only mainland-China-based director. During this period, he also started to learn about batteries and met businessmen with international connections. One was T.H. Chen, a Taiwanese who had been born in Hunan province, had a PhD in physical chemistry from the University of California at Berkeley and had previously worked for IBM. By the end of the 1990s Zeng was planning to leave to move to the larger nearby city of Shenzhen, which borders Hong Kong. But the CEO of the magnetic hard drive company, Liang Shaokang, persuaded Zeng to start a battery company. Initially reluctant, Zeng agreed and in 1999 ATL was formed in Hong Kong with T.H. Chen becoming the chief executive.

ATL aimed to make batteries for mobile electronics. It was ideal timing: mobile phone sales were growing and many were connecting to the internet, which required more portable power. In late 1999 Nokia launched its 7110 phone which could access information over a mobile wireless network. And in Japan the internet-connected i-mode phone was also launched by NTT Docomo in the same year. By 2002, a total of ninety-five percent of mobile phones used lithium-ion batteries.[7] Sales of portable MP3 players were also growing, and accelerated after the 2001 launch of Apple's iPod. The lithium boom had begun and Dongguan became a centre for producing mobile phones, chargers and accessories.

When it started, however, ATL had little of its own intellectual property or any breakthrough technology. Zeng and his colleagues spent $1 million buying a lithium polymer patent from Bell Labs in the US. But when they returned home ATL found making the technology work was not as easy as they had thought – the battery expanded when it was repeatedly charged and was also at risk of exploding. As they struggled in Dongguan they worried it might spell the end of the young company. They spent two weeks working overtime to overcome the difficulties by trying different electrolyte combinations. Finally, they got the lithium polymer battery to work. Once they'd done that, they managed to rapidly cut the cost of production – a model Zeng would repeat later with electric car batteries. ATL managed to produce batteries at half the cost compared to their Korean competitors. Their lithium polymer battery was also thinner than other models and could be shaped according to the device. ATL was profitable within three months of the battery's production.

It was the beginning of China's move into batteries – a business dominated by Japan since Sony commercialised the first lithium-ion battery in 1991. China had come slowly to the battery revolution, with its first lithium battery developed at the Institute of Physics in 1995,[8] four years after Sony had commercialised the technology. Even by the year 2000 Japan accounted for ninety percent of the world's lithium-ion battery production with 500 million batteries, and China only producing 35 million.[9] Yet by 2001 ATL shipped over one million batteries and its batteries were used in Bluetooth headsets and portable DVD players. The same year China joined the World Trade Organization, which opened up the country to significant foreign investment. ATL helped China become a high-value producer of batteries.

At the time Sony supplied batteries to Nokia and Chinese rival BYD batteries to Motorola.[10] But ATL quickly became the choice for other phone makers. In 2004 it entered Apple's supply chain, providing batteries to its iPod. On T.H. Chen's visits to the US to see his family his son remembered him showing 'a new gadget and saying in a matter-of-fact way: "we made the battery for this". The technology was a little beyond me, but I was still scheming on how I could leverage his connections to acquire my own iPod,' he remembered.[11] A year earlier ATL received $30 million in investment from top-tier US private equity company the Carlyle Group as well as 3i Group from the UK.[12] 'ATL is a notable example of next generation Chinese businesses making a mark in the global technology arena,' Carlyle said. 'Its polymer batteries are thinner, more flexible and safer than traditional lithium-ion batteries, at the same time offering consumers a 10–20% higher energy density.'[13] Gabriel Li, managing director of Carlyle Asia Technology Fund, noted that Dongguan was 'the lowest manufacturing cost location in the world', from which ATL sold batteries to leading multinationals.[14] It was all the benefits of globalisation in a nutshell for China: foreign technology, foreign money and a Chinese company producing high-value products. A year later Japan's TDK bought the whole of ATL for $100 million.

Tokyo-listed TDK was a maker of electronic components that had been founded in 1935 to develop a ferrite magnetic material discovered a few years earlier by Dr Takeshi Takei, a professor at the Tokyo Institute of Technology. The ferrite material was initially used in radios, helping to reduce their weight and improve the sound quality. Ferrite from TDK was also used in the cathode ray tubes of television sets after the Second World War. The company had a deep and broad experience in consumer electronics, from cassette tapes to videotapes. It was a giant and Zeng could have had a comfortable executive career there.

But by the early 2010s Zeng could see that China's government was serious about supporting an electric car industry – both to clean up air pollution and reduce reliance on oil imports, but also to attempt to leapfrog Western carmakers and have a stake in the car industry of the future. For much of the twentieth century China had been absent from the development of the internal combustion engine. In the international heyday of Shanghai in the 1930s it was Buicks and Fords that cruised the tree-lined streets, not Chinese-made cars. And China's thirty-year car boom had generated fortunes for Western carmakers and their joint-venture partners in China, but there were still no real Chinese global car brands. Few people on the streets of London or New York could name a Chinese car company, let alone long to buy a Chinese-made car. At the same time China's car boom had been a tragedy for the global environment, even more so since it started after the world was fully aware of the dangers of climate change and the contribution of man-made emissions. The legacy was a thick layer of pollution in cities like Beijing and across China, whose human cost was borne out in the number of premature deaths. As more people moved to China's cities the number of cars increased, creating a toxic combination. By 2020 China was importing more oil than any country in history, at more than thirteen million barrels a day.

The solution was electric cars. But how to get people to buy them? Starting in 2009, Beijing began paying people to buy electrics, using a mixture of subsidies and incentives. It was a government intervention like no other. One man was pivotal in this regard – a former Audi engineer who had returned to China to help his homeland and whose ambition propelled him to an influential position within the Communist Party. Wan Gang had been sent down to the countryside during the Cultural Revolution, and there he learned about engines by taking apart a village tractor. He later studied for a PhD in engineering from Clausthal University of Technology in Germany and then spent a decade as part of Audi's R&D team, where he worked on fuel cells. He returned to China in 2000 to head work on Tongji University's programme in hydrogen fuel cells, becoming the president of the university only four years later. His rapid rise did not stop there: three years after that he became head of the Ministry of Science and Technology. Wan was a 'sea turtle', a returned Chinese, and unusual in not being a member of the Communist Party. Yet he was determined for China to move ahead in electric cars and compete with Japan. It was a message that was well received by the Party and China's technocratic leaders, who had long wanted to catch up in science and technology. Electric cars had been included in China's '863 program', a National High-Tech R&D Program launched in 2001.

In the midst of the global financial crisis, hope was growing again globally for electric cars despite many false starts. The industry was seen as a potential salvation for the slowdown in economic growth. Foreign observers waxed lyrical about China's BYD, which had grown from a battery maker like ATL into an electric car producer. In late 2008 Warren Buffett had bought a ten percent stake in the company. But the truth was that few consumers were interested in buying electric cars. Hybrids, which used an engine as well as an electric motor, seemed like a better bet, and the Toyota Prius was going from strength to strength. If the market didn't exist, why not just create it? Like most new programmes in China, Beijing began with a series of pilots. In 2009 Beijing launched a pilot programme in ten cities to subsidise electric buses and public vehicles. This was expanded a year later to cover private cars in six cities, where local governments could also provide their own incentives. Subsidies were generous: in Hangzhou consumers received nearly $20,000 from central and local governments to buy an EV.[15] It was the beginning of a lavish subsidy machine that helped launch thousands of electric car start-ups. Local governments vied with each other to buy electric taxi fleets or buses. Sales of China's electric buses soared and China accounted for almost all the global deployment of electric buses by the end of 2018, with 421,000 e-buses on the roads (compared to 425,000 globally).[16] Between 2009 and 2017 the Chinese government spent an estimated $60 billion on subsidies. It was an unprecedented – and expensive – industrial intervention that quickly made China's electric car industry the biggest in the world. Underlying the growth, however, were purchases by China's local governments, which accounted for the bulk of sales between 2013 and 2016, according to the scholar Ilaria Mazzocco. Local governments became innovative in helping to finance fleet purchases – sometimes too innovative, as cases of subsidy fraud proved.

'Subsidies at the local level played a starring role in creating both public procurement and private passenger markets,' she wrote.[17]

It was in this environment that CATL, whose Chinese name means 'the age of Ningde', was split off from ATL, with the support of the Chinese government, in order to create a fully Chinese battery champion. (ATL initially held a fifteen percent stake in the company but sold this in 2015.) Zeng hired Bob Galyen, a six-foot-six blonde American who had worked in batteries for thirty years, including on General Motors' doomed EV1 project in the late 1990s, as chief technology officer. Galyen moved to Ningde without his family to work full time for the company. He was welcomed warmly, he recalled, despite standing out. 'I'm 6'6", a bit of a giant, let alone in a back city like Ningde ... There were no stop lights or stop signs, it was the wild wild west,' he recalled. 'By the time I left everything was modernised.' One of CATL's earliest deals was with BMW's Chinese joint venture, BMW Brilliance. It was transformational: the rigorous testing of BMW helped improve CATL's standards, and BMW maintained a permanent floating cohort of staff checking on every step of the battery manufacturing process.[18] 'BMW's reputation in the industry has helped lift CATL out of obscurity and into stardom,' the state-owned *China Daily* said.[19] China had put pressure on BMW to help its joint venture develop EVs, and in 2013 it launched the Zinoro 1E, an electric SUV that used CATL batteries, with a range of 150 kilometres. 'We have learned a lot from BMW, and now we have become one of the top battery manufacturers globally,' Zeng said. 'The high standards and demands from BMW have helped us to grow fast.'[20] Between 2014 and 2017 CATL's sales increased at a compound annual growth rate of 263 percent. In 2016, according to Galyen, CATL delivered more battery packs to battery company Yutong Bus than Tesla had used in all of its cars since it began making them.[21] In 2017 CATL filed for an initial public offering (IPO) on the Shenzhen Stock Exchange, with the help of Goldman Sachs. The company raised $853 million and became the world's largest producer of electric car batteries with a fifty percent share of the Chinese market. It would maintain that position consistently for the next four years.

How had it managed to grow so fast? The truth was that Zeng had already helped build ATL into a globally successful battery producer. He and his colleagues had considerable knowledge and skill from the competitive global mobile phone market. He took that manufacturing experience to CATL, where the company worked with machine manufacturers to design its own machines to make a high volume of batteries. Once the staff were educated in how to use them, this allowed for a higher throughput of batteries of a higher quality than other manufacturers, Galyen said. He called this magic combination 'man and machine' or 'M&Ms'. 'He [Zeng] already knew how to make batteries,' he said. 'All he had to do was make them bigger.' Yet the other crucial factor was brute protectionism from Beijing. China's subsidy machine was specifically designed to boost domestic companies and create a Chinese electric car ecosystem. Between 2016 and 2018 the Ministry of Industry and Information Technology produced an annual list of approved EV battery suppliers – all of which were Chinese companies. 'The premise is that locally produced cars in China are obligated to use local batteries,' Jochem Heizmann, chief executive of Volkswagen Group China, said.[22] Introduced in the world's largest electric car market just as sales were taking off, the policy was extremely powerful – effectively barring Korean battery companies LG Chem and Samsung SDI from competing. 'It was actually a brilliant strategy,' Jim Greenberger, founder of NAATBatt, the North American trade association for advanced battery technology, told me. 'CATL was the winner and they have used that scale to compete very effectively in the export market. That's the issue we're dealing with in the west: how to compete with Chinese companies that have gotten to scale, via use of industrial policy.'

CATL under Zeng was far from complacent, however. He saw the battery industry as akin to a war. 'We are competing with gasoline cars,' he said. 'If we can't win against gasoline cars, there's no place for us in the market.'[23] He ploughed money into R&D, spending more than his rivals. By 2019 CATL had over 2,000 patents in batteries and battery charging. CATL modelled itself on the hard-working culture and R&D focus of telecommunications giant Huawei, whose rise had led to growing political concerns in the US. Zeng knew that CATL could not always rely on Chinese subsidies for help. In an internal email in 2017 he referred to a Chinese allegorical saying: 'the typhoon is coming, even the pigs can fly', which suggested that with government support any company can do well. But Zeng was not convinced. 'But are the pigs flying? Once the typhoon passes, what is the situation for the pigs that are left?' he asked. Neill Yang, head of marketing at CATL, put it more explicitly: 'People think we're a big successful company, but we think we're in jeopardy every day. The market environment and technology changes so fast that if we don't follow the trend we could die in three months.'[24] CATL relentlessly pursued innovations that helped push down the costs of batteries. It combined battery cells directly with aluminium battery packs and removed the battery modules, which saved materials and reduced the battery's weight. In 2020 it also became the first battery maker to announce it had developed a 'million-mile' battery that would last for sixteen years. It was a significant achievement as it meant batteries could outlast the cars, and could be re-used, helping to push costs down even further. When Tesla built a factory in Shanghai in early 2020 it chose CATL as a provider – the company met both its cost and its technology requirements. The state-owned *China Daily* was metaphorically effusive over CATL: 'in only a few years [CATL] has grown from an ordinary fish in the hatchery of China's new energy vehicle market into a veritable sperm whale'.[25]

China's manufacturing prowess – in the form of CATL – had helped to significantly drive down the global cost of batteries, making electric cars increasingly cost-competitive against petrol cars. It was a similar process to what had happened in almost every other clean technology, from solar panels to polysilicon production. Subsidies had led to a rush to produce and overcapacity, but had ended up with China dominating global markets. The cost of lithium-ion battery packs fell by eighty-nine percent in real terms between 2010 and 2020 – going from $1,100 per kilowatt-hour (kWh) in 2010, to $137 per kWh in 2020, according to *Bloomberg New Energy Finance*.[26] CATL did an 'outstanding job of taking an existing technology and scaling it – taking it from low volumes to high volumes in a very cost-effective and high-quality way', Galyen, CATL's former chief technology officer, said.[27] It would be hard for other countries and companies to catch up unless they invented a new battery chemistry, he added.

By 2021 CATL was the second-largest listed company in China with a market capitalisation of over $200 billion. Zeng was keenly aware of the importance to China of having a strong battery and car industry. New technologies were closely connected to national power and strength, he said. Before the nineteenth century China had been a leader in the global economy thanks to ironware and the 'four great inventions': papermaking, printing, gunpowder and the compass. But the invention of the steam engine in the eighteenth century had caused the UK to become a global factory and China to 'miss an opportunity, and to decline from a peak'.[28] Nevertheless, in his view, over the last forty years China had made some advancements in certain areas such as 5G technology. The current global economy was focused on renewable energy, electric cars, biotechnology and artificial intelligence, he said. The change in the car industry was an opportunity not seen in a hundred years. The battery industry could make China a 'strong country', economically and politically. 'In the whole of human history, every period of new technology represents an advance in productivity. The appearance of new technology not only encourages break-through economic development, but also changes the structure of global competition,' Zeng wrote.[29]

However, it was not only the battery industry that China had dominated, but also the raw materials needed for the manufac-turing process – and that required China to look outside its own borders.

5

The Chinese Lithium Rush

'Lithium is extremely common on Earth. Found almost everywhere.'

Elon Musk, Tesla CEO[1]

On a cold autumn evening in late 2019 Wang Xiaoshen entered a private room at the Royal China Club in London's Marylebone to celebrate his latest acquisition. As the large dishes of seabass in spicy Sichuan sauce and smoked duck arrived and bowl-shaped glasses of red wine were served, the conversation turned to the depressed lithium prices that had prevailed for the past year. A year earlier, in 2018, Wang's company Ganfeng Lithium had listed on the Hong Kong Stock Exchange with the help of Citibank, but had priced its IPO at the bottom of the expected range, raising $422 million, instead of the $1 billion it had hoped for. On its first day of trading the company's shares fell by twenty-nine percent, leading reporters in the city, where retail investors normally lined up to buy into new mainland companies, to call the IPO 'dreadful'.[2] No one much cared for lithium, it seemed, even as sales of electric vehicles were accelerating. Investors viewed it as an unloved commodity – it was abundant in the earth and could be dug up in Australia and put on a boat to China easily. It was an inauspicious start for the company's entrance to the global markets from its home in the central Chinese city of Xinyu.

But Wang was unfazed. Instead of retrenching, he saw the weak market as the opportunity to expand. The dinner had been arranged to celebrate his purchase of a stake in Bacanora Lithium, a company with a project in Mexico, led by an Englishman with a full stock of white hair, Peter Secker, who sat opposite Wang at the table and remained mostly quiet. Bacanora Lithium was listed on the UK's Alternative Investment Market for small companies, and Secker, an experienced mining engineer, had tried his best to increase interest in the stock. In 2015 the company had announced a deal with Tesla, which sent Bacanora's shares soaring, even though the deal was conditional and later quietly fell apart. Earlier that year Secker had been forced to abandon a planned $100 million fundraising the morning it was supposed to close, shocking investors. There was simply no demand. Secker had now handed the keys to the project to Wang and thrown in his lot with the Chinese. To Wang the project was perfect; it was one hundred miles from the US border so it would allow easy access to the US market without alerting the authorities to the presence of a Chinese company on its home turf. 'The US is off limits to us,' Wang conceded, due to the political pressure mounting in the country under President Donald Trump, who had launched a trade war with China.

Yet America produced virtually zero lithium.

Wang was uniquely relaxed, warm and comfortable in his own skin. He exuded a quiet confidence and a light sense of humour. He had managed to learn fluent English despite never having lived overseas. He was the picture of a successful Chinese executive: he wore a smart black watch, drove a Tesla Model X and had a child studying in New York. 'Wang Xiaoshen thinks about lithium 24 hours a day,' his friend Joe Lowry told me. Yet beneath his genial exterior Wang was obsessively determined: in the space of ten years, he had built his company into one of the biggest lithium producers in the world. 'They very much follow the framework of Faster, Better, Cheaper,' John Kanellitsas, Wang's partner on a project high up in the mountains of Argentina, told me. 'Maybe add Bigger as well.'

*

A silvery-white metal that reacts violently with water and is so volatile that it is only found in compounds in nature, lithium was discovered by Johan August Arfwedson, a Swedish chemist, in 1817. The lightest of all metals, it is one of the most abundant in the earth's crust and is also found in oceanwater. Most of the metal that we use today may have been formed by nuclear reactions that cause stars to explode, which then distributed lithium throughout the galaxy.[3] An element that scarcely any of us can picture, lithium is essential to our digital world and the green economy. We rarely see it or feel it, and there are few mines in much of the Western world, but we carry lithium with us 24/7, in our mobile phones, smart watches and tablets. And without lithium, there would be no electric vehicle, and no Tesla. Just as bronze, iron, coal and oil defined different epochs of our history, lithium will be the element that shapes our future.

One of lithium's key early uses was in treating gout, but it was soon discovered to have an uncanny ability to improve people's moods. In the early twentieth century lithium was included in drinks such as 7 Up (whose original name was Bib-Label Lithiated Lemon-Lime Soda) and natural mineral springs that contained lithium such as the Lithia Springs in Georgia became popular tourist spots, visited by Mark Twain and President Theodore Roosevelt among others.[4] The big breakthrough for clinical usage came in 1949, when Australian psychiatrist John Cade showed how lithium could help treat patients suffering from bipolar disorder. Working in an abandoned pantry near Melbourne, Cade, who had been interned during the Second World War in the Japanese camp of Changi in Singapore, collected urine samples from mentally ill patients and then injected them into guinea pigs. He found that lithium reduced the toxicity of the urine, but also seemed to improve the mood of the animals.[5] His discovery enabled lithium's key treatment for bipolar disorder. It was the first time a drug had been discovered to deal with mental illness.

The first industrial production of lithium was in 1923 by the German company Metallgesellschaft in Langelsheim, using a type of lithium-bearing mineral called zinnwaldite.[6] Soon after, the Maywood Chemical Company in New Jersey started production, followed by Foote Mineral Company and Lithium Corporation of America. Lithium found an early military use in submarines, as an absorber of carbon dioxide. Later it was critical to the development of nuclear weapons and the hydrogen bomb.

For batteries, scientists have not yet found a better alternative. That's due to lithium's light weight (especially compared to lead used in lead-acid batteries) and its electrochemical potential. Electricity is simply the movement of electrons, the charged particles surrounding the central nucleus in an atom. Lithium readily loses an electron, forming a positively charged ion, and easily burrows into host material such as the graphite in the anode and the cobalt oxide in the cathode, allowing the lithium to move backwards and forwards during discharging and recharging. Because lithium is so light it makes up a small amount of the weight of a battery, which is critical since heavier batteries mean less range.

Lithium's abundance is good news for our efforts to combat climate change since we will need lots of it: for electric vehicles as well as for storage of renewable energy for our power grids. In theory we could run every car on lithium batteries for a billion years, with the amount of lithium in the earth's crust. But that's not realistic: lithium still needs to be extracted from the earth economically, which requires someone to finance and build a mine. It also needs to be carefully processed into a form used for batteries

– lithium carbonate or lithium hydroxide – without impurities that could cause a battery to catch fire.

For most of the twentieth century the US was the largest producer of lithium. The global market was dominated by two companies: Philadelphia-based FMC and Foote Mineral (later Cyprus Foote Mineral and then Chemetall SA), which both owned mines in North Carolina. FMC had started life as a company making insecticide spray pumps to combat an infestation damaging California's orchards in the late nineteenth century. In 1985 it acquired the Lithium Corp of America (known as Lithco), the world's largest producer. FMC supplied lithium to Sony for the first commercial lithium-ion batteries in its camcorders. At the time the US was a net exporter of lithium, from the hard-rock mines in North Carolina. But then in the early 1990s after a failed bid to develop lithium in Bolivia, Lithco moved into Argentina, where it produced lithium from high up in the Andes, evaporating it out of naturally occurring brine using large ponds rather than digging it from rock. The method was much cheaper since it relied on sunlight to do most of the work. It was the same process being deployed across the border in Chile by Foote Mineral and separately by a Chilean company, Sociedad Química y Minera de Chile, which was owned by the son-in-law of the country's dictator Augusto Pinochet. SQM, as it was known, would join the two American companies as a global producer. In 1998 FMC shut down its higher-cost mines in the US – ending America's largest source of domestic supply, leaving only one project in the country, the Silver Peaks mine in Nevada. The mineral-rich lithium veins that stretched through North and South Carolina would remain untouched for decades (Foote Mineral had shut its Kings Mountain mine earlier in the 1980s). While FMC remained the dominant supplier of lithium chemicals to the battery business, for mobile phones, camcorders and personal computers, the lithium came from Argentina and was processed in chemical plants in Bessemer City in North Carolina.

By the turn of the century the lithium industry was an oligopoly made up of the 'Big Three' producers. It was not really an industry that was much in the public eye. It was a niche market that served dull uses such as the making of glasses and grease, which hardly excited investors. While smartphone sales had doubled lithium demand since the beginning of the century, it was from a small base. In 2000, China did not even feature as a meaningful market in a budget presentation for FMC, according to Joe Lowry, who handled lithium sales at the company. A few years later when news came out about Nissan's Leaf electric vehicle and Chevrolet's Volt the lithium industry reacted cautiously. Electric vehicles had failed so many times before. The industry's expansion plans were limited. 'The "Big 3" were self-satisfied – fat, dumb and happy with the world as they knew it. They enjoyed this new battery demand but were unsure where it was going and were slow to change their processes to accommodate it,' Lowry recalled. 'It was in their interest to have an expanding market but they were very conservative about that subject,' James Calaway, who helped start a new lithium project in Argentina, remembered. 'They acknowledged it and they knew there was something there but they would have assigned high risk to it. They didn't do anything to react to it. And I think they misjudged that terribly. They had an opportunity at that time to make it much more difficult for new entrants.'

It was a cosy market, ripe for disruption.

*

On a drizzly, cold March day, I'm driven in a gleaming black Tesla Model S through the wet paddy fields outside an industrial city in the middle of China to see the beating heart of the battery economy. 'This all used to be countryside,' a young staff member, who studied in Bristol, told me as we approached the factory gates. Nearby, cows strolled freely despite the fumes emanating from the plant, which rose out of the cold mist, a concrete hulk of pillars, encircled with large rusted pipes. Inside, grey ground rock – mined in Australia, shipped down the Yangtze River and then trucked to the plant – was heated to 1,000°C in a giant coal-fired boiler. The rock, which contained around six percent lithium, was then leached with acid, dried, and purified into a fine white powder that would then be mixed with other metals to make a lithium-ion battery. Ganfeng was China's largest producer of lithium hydroxide, a high purity product that was used in Tesla's more powerful batteries. It sold for around $9 a kilogram when I visited. I was shown around the factory by a Ganfeng worker in a windowless buggy, despite the wet cold. 'When they bought it, it was summer,' he joked. There was no heating in China south of the Yangtze, due to government regulations, so the buildings were cold inside too. There were around 1,000 workers at the site and the associated laboratory that tested the lithium down to the atomic level to make sure it met carmakers' specifications, he told me. The plant ran twenty-four hours a day, using coal-fired power. (The company needed about two tonnes of coal to produce a tonne of lithium.) The final bags of white lithium, big as bales of hay, were then trucked back to the Yangtze River to be shipped to customers, and turned into batteries. By that time the lithium had already travelled thousands of miles on its journey from mine to car.

I had arrived in Xinyu on the high-speed train from Shanghai, travelling through a wet grey-green landscape of sloping hills and paddy fields. The city, known for its steel industry, looked like many Chinese cities do on arrival. There was the stadium, built by the local government, despite not having a major local football team. There was the requisite vast station building, and sprouting up all around it the forests of concrete and pastel-coloured new apartment blocks. I remembered the city as the home of LDK Solar, a New York-listed solar manufacturer that the local government and China's state-owned policy bank China Development Bank had heavily supported despite mounting competition. In 2012 the local government bailed the company out to the tune of $80 million, but it still collapsed into bankruptcy a few years later. It served as a reminder of how often China's state-backed projects went wrong, despite the image people had overseas of a strategic masterplan. As a result, most Chinese entrepreneurs were paranoid about competition. In lithium, Xinyu hoped for a better result.

Thanks to Ganfeng the city had become a vital node on the global electric car supply chain – the twenty-first-century equivalent of the refineries, pipelines and ships that supported the age of oil-based transport. Ganfeng, along with its Chinese rival Tianqi Lithium, had built China's dominance in the global battery supply chain. While China only mined a fraction of the world's lithium, it processed over eighty percent into batteries, compared to one percent for the US. It was this step in the supply chain that was so important. China was not blessed with large deposits of battery minerals – lithium, cobalt, nickel, graphite and manganese. But it had captured the processing step, meaning that no matter where lithium was dug up, it had to go to China. China had become the central clearing house for the mineral. There was simply no alternative. It was a strategic success for China: once the lithium reached China, it was more likely to be turned into battery materials by a Chinese company and end up in a Chinese-made battery. Tesla's entry into China in 2019 as the first foreign carmaker to build a plant in the country had been premised on its use of domestic, Chinese suppliers, one of which was Ganfeng.

At dinner on my first night in Xinyu I asked Anna Liao, a Ganfeng employee who was showing me around, whether the West would ever catch up. 'They have missed out on the first wave,' she said. Ganfeng was integrated all the way from the raw materials that came out of the mine to the final battery. It also recycled old lithium-ion batteries to turn into fresh raw materials. It would be hard for many Western companies to compete with the company's costs. The following day I donned a bright yellow hat and gown and passed through an airlock to see the company's battery factory. Battery materials were coated onto thin films of copper by large rolling machines behind glass walls. The manufacturing line hummed with efficiency; the only movement was that of the Kuka robots (a German robotic company that had also been acquired by a Chinese company a few years earlier). It was the opposite of the labour-intensive factories one imagined in China. The batteries ended up powering China's vast fleet of electric buses.

Ganfeng was not content just to make lithium for the world. It was also researching future battery materials that would enable electric cars to be charged in minutes, cementing their competitiveness against petrol. On my final day in Xinyu I travelled with the Ganfeng staff to the nearby town of Yichun, the birthplace of the famous ancient Chinese poet Tao Yuanming. Inside a nondescript brick building workers handled rectangles of pure lithium using gloves inserted into an airtight container full of argon. The metal was then tightly packed into airtight aluminium foil bags for shipping to customers. This was lithium metal – a highly volatile substance, but one that allowed much more energy to be stored in a battery. Lithium metal could store up to ten times more energy than the graphite (a form of carbon) that was used in most batteries. Such an increase in energy density would also open up other applications to batteries such as aeroplanes. It was the same material that Nobel laureate Stanley Whittingham had tested while at Exxon Mobil in the 1970s, when he built the first lithium-ion battery. Long a holy grail of battery scientists, its usage was now being perfected in the middle of China. Wang told me he expected the technology to be in cars by 2025 or 2026. If Ganfeng could succeed it would put the company at the cutting edge of the global battery market and entrench even further China's success in electric cars.

The prospect of an electric vehicle revolution was unimaginable for Wang Xiaoshen growing up. He was born in China's far Western Xinjiang region, where his grandfather had gone to help on an irrigation project in the desert, sent in 1943 by Chiang Kai-shek's Nationalist government in war-torn Chongqing. The region was barely under the control of the Nationalists, who were busy fighting the Japanese, and its resources had been exploited for years by the Soviet Union. Wang's mother's family had moved to Xinjiang. After graduation in 1990 Wang worked for one of the country's first lithium plants in the regional capital of Urumqi, which had been built in 1958 to supply an isotope of lithium to China's nuclear weapons programme. 'I had no idea about lithium,' Wang remembered. The '115 Factory' processed lithium rock from the remote Altay Mountains, where Soviet geologists had discovered rich sources of metals in the 1930s. By the late 1940s, with the help of Soviet equipment, the mines were producing 1,000 tonnes a year of minerals, including beryllium and tungsten. The minerals were all exported to the Soviet Union, despite repeated objections from the Chinese that it amounted to illegal mining on Chinese territory. Control returned to China after the ascension to power of the Chinese Communist Party in 1949, when the resources were taken over by the state-owned Xinjiang Non-Ferrous Metals Company, where Wang would later work. The Communist Party viewed Xinjiang as a barren land that was full of a 'limitless supply of treasures'.[7] At first, the mines in Xinjiang's far north continued to ship resources to the Soviet Union to repay loans, but China's nuclear programme created a source of domestic demand for lithium.

As China opened up in the 1980s, the state-owned plant in Urumqi started to supply civilian applications such as glass and aluminium production, and air conditioners. It became the country's biggest lithium supplier. But safety standards at the factory were poor, and dust mixed with the chemicals to burn the skin. Lithium is one of the most reactive metals whose impact on those who suffer repeated exposure is not well studied. Wang's abiding memory was of dirt getting everywhere – on his clothes, his shoes, his hands. When he entered the plant to produce lithium hydroxide he immediately felt the dust in his mouth. 'Everywhere was very dusty and noisy,' he recalled. 'At that time the protection was not that good. The dust could go anywhere; when you sweat you felt burned.' The burns left scars on his skin. Outside, pollution from the coal-fired power stations in Urumqi turned the snow black in winter. One day Wang saw a colleague die in an accident because he was too short to operate the equipment and had forgotten to stop the centrifuge. 'That's a tragedy,' Wang recalled from his office high up in a Shanghai skyscraper.

By the early 1990s Xinjiang's lithium resources had run out, and the plant needed to import lithium from Australia, from a large mine known as Sons of Gwalia (the Greenbushes mine, the largest lithium mine in the world; see the next chapter). To make matters worse, in 1997 SQM of Chile started to produce lithium cheaply using the sunlight from high up in the Atacama Desert, and began exporting it, which crashed prices in the global market. Soon, the Xinjiang plant would be bankrupt. It was not good timing: demand for lithium was beginning to increase due to the rise of mobile phones. Wang moved to Beijing to set up the company's office there and then left the company in 2002. 'I said no more.' He went to the eastern city of Suzhou to work in a power tools business that was looking to make lawnmowers with lithium batteries. The country's economy was growing by double digits, yet it was burning more coal and importing more oil than ever before. Pollution darkened the skies and fouled the waters. Sales of petrol cars were accelerating. China's start in lithium did not look promising.

But an entrepreneur based in the central province of Jiangxi set out to change that.

One of the interesting features of China's prominence in lithium was that it had not been achieved by the state-owned companies, as many people outside China thought. Instead, it was the product of a group of entrepreneurs who 'jumped into the sea' from the state sector, as the Chinese call the launch of a private business – sensing an opportunity in batteries for mobile phones and then electric cars. Li Liangbin, who founded Ganfeng in 2000, had previously worked for Xinyu's state-owned lithium plant, which had supplied lithium to China's nuclear weapons industry in the 1960s, from local lithium mines. As a young man he had been given a choice of joining the large state-owned steel plants with their thousands of workers or the lithium plant, which had less than 1,000 workers, and he thought the latter would provide more money since there were fewer workers. He rose to be a director at the plant and believed he had a bright future. Yet just like Wang's factory in Xinjiang, the Jiangxi factory could not compete against the cheap lithium being exported from Chile in the 1990s. In 1997 Li left his job, and decided to form Ganfeng. At first, he bought lithium supplies from Xinjiang Non-Ferrous, which is how he met Wang. He then convinced him to join Ganfeng in 2006.

Ganfeng was from the very beginning supported by the 'Big Three' lithium producers. The company started as a customer of SQM, as well as America's FMC, buying lithium that it would then process in China. In 2004 Li flew to Chile and even offered a fifteen percent stake in Ganfeng to SQM, in order to secure lithium supplies. 'At that time SQM hired Deloitte to do due diligence and after that [SQM] quit. It said "no thank you". Ganfeng at that time was a very small company,' Wang said. A few years later SQM returned and offered to buy the company, after its chairman Julio Ponce Lerou (who we'll learn more about in the next chapter) visited Xinyu in 2007. 'They understood we have the capability, they changed their mind,' Wang recalled. By this time FMC and Rockwood (later Albemarle) were also interested in a takeover. Wang thought hard about it but at the same time he was being approached by Chinese investment funds, including one run by state-owned mining giant Minmetals, which urged him to list on the domestic stock exchange instead. So Ganfeng became a competitor, listing on the Shenzhen Stock Exchange in 2010. SQM soon forbade it from buying any more of its lithium. If the company had bought the stake it had been offered, it would be worth around $5 billion today.

At the time China was dotted with small lithium converters who all competed to find raw lithium resources to process into chemicals. It was easy to get bank credit and there was little environmental scrutiny, with plants looking like something out of a Charles Dickens novel in the early 2000s, according to Lowry, who travelled to China regularly while working for FMC from Japan and then Shanghai. When he visited Ganfeng at the time, he saw workers wearing shoes with no socks and wearing rubber gloves to handle a cauldron of lithium chloride in the August heat. 'In North Carolina we did it in a climate-controlled building and the guy wore a space suit,' he added.

With China's famous Xinjiang lithium deposits long exhausted, and few other large economically viable deposits, Ganfeng had no choice but to find resources outside China's borders. Wang began to look everywhere – even sending a team in 2013 to the Irish town of Carlow outside Dublin to look for the mineral. In the shadow of a castle, geologists drilled for lithium. It had better luck a few years later in Australia, a country rich in lithium and home to the Greenbushes mine. In 2015 Ganfeng bought a twenty-five percent stake in the Mount Marion mine in Western Australia, outside the gold mining town of Kalgoorlie. It was a move that fired the starting gun on a new gold rush that would see Australia become the dominant lithium producer. A country long famed for exporting its coal and iron ore, and dragging its feet on climate action, was becoming a key player in the clean energy supply chain.

*

Over 1,500 kilometres from Perth, giant mines scar the red, Mars-like landscape in the empty Pilbara region. Double the size of France this is 'fly in, fly out' country, known as FIFO, as miners commute from Perth to the mines for weekly shifts before returning home to their families. To the first Dutch explorers in the seventeenth century the area was worthless, but the rich resources of iron ore beneath the earth have since enriched Australia, making it one of the few countries not to have a recession for more than a quarter of a century. Since China overtook Japan as the biggest buyer of Australia's iron ore in 2005, its demand has been insatiable. Iron ore, the rock that contains naturally occurring iron, is a key ingredient for steel production, along with coal. And putting steel rods into concrete makes a perfect construction material. As China built ever more apartment buildings and new cities at the beginning of the twentieth century its demand led to an upswing in prices that was dubbed a commodity 'supercycle'.* China turned into one giant construction site that needed feeding with raw materials. In turn, Australia became one giant mine, digging up and shipping minerals to China. With iron ore prices sky-high, any amount of spending justified getting another tonne out of

---

* While there is no agreed definition of a supercycle, it has been commonly used to describe a period where commodity prices rise above their long-term trend for between ten and thirty-five years.

the ground, and so spending ballooned. Australian miners became rich and the two countries became locked in an uneasy embrace.

But in 2014, the music stopped. The post-financial crisis boom in Chinese bank lending and construction had tapered off, and commodity prices crashed. China's huge appetite for Australia's iron ore and coal seemed to be waning. Global mining companies were left drowning in debt and fending off attacks of hedge fund investors. Prices for iron ore slumped by more than forty percent, putting pressure on Australia's smaller producers, which had higher costs of production. Mine workers who had lived the high life in Perth and other cities started to feel the pinch and house prices fell. Analysts started to talk about the end of the Chinese miracle, and the shift to a more consumption-based economy. Yet the crash was also a chance for Australia to reassess and focus on the future. Coal miners, and iron ore producers, who had given no thought to helping the environment and driving electric vehicles, began to look for the next big thing.

The same year five geologists packed their bags and headed out to the Pilbara to look for fresh riches. The friends, who had met at university in Perth, began looking for tantalum, a metal used in smartphones. But they found something else instead – a massive deposit of lithium, sandwiched in between the iron ore mines where lithium-rich rocks called spodumene could be picked up from the red dusty ground. Unlike in Chile, where lithium is evaporated from lithium-rich brines, in Australia it is found in rock that is dug up and crushed like copper and other metals. Neil Biddle, one of the geologists, who came from a family of prospectors, did not know much about lithium. 'It was not a mineral that was even in our sights,' he recalled. At first, the geologists thought the creamy grey spodumene rock they found was feldspar, an abundant rock that makes up more than half of the earth's crust. 'Even though there was plenty of spodumene at surface cropping out we didn't recognise it.' But he quickly became interested in electric vehicles, and took to wearing a Tesla-branded cap. That was unusual in Australia – the country had few electric cars on the road. His interest increased after Tesla announced in September 2014 it would build a giant battery 'Gigafactory' outside Reno in Nevada. 'It became very obvious very quickly that Pilgangoora was a big lithium deposit. Everywhere we drilled we were getting lithium.'

A few years later at the opening of the Pilgangoora lithium mine guests dined on fresh prawns, oysters and roast beef in the 40°C heat. The new chief executive of Pilbara Minerals, Ken Brinsden, could still see the iron ore mine where he used to work when he stood on a ridge of the new lithium mine. An affable man with clear pale-blue eyes and boyish blonde hair, he had spent a decade in the Pilbara, riding the boom and then bust in iron ore. Brinsden had grown up in Sydney but his parents were from Kalgoorlie in Western Australia, a centre for gold mining. His family had long links to the mining industry, starting with his great grandfather, who was a contemporary of former US president Herbert Hoover in the Gwalia gold mine in Western Australia. Hoover had gone to the goldfields in his early twenties and helped turn around the Gwalia mine, making the fortune that would help him become president.[8] Brinsden continued the family tradition and studied at the Western Australian School of Mines at Curtin University in the early 1990s before going to work for a number of gold mines. He joined iron ore miner Atlas Iron in 2006, just as the Chinese-driven commodity supercycle was getting started. In 2015 he was introduced to Biddle who told him that lithium was interesting and that the market had 'got to be heading in the right direction'. It has got to be better than iron ore, Brinsden thought. At the time he joined Pilbara it was a penny stock. But he needed to find financing to build the mine.

Then he went to visit China.

Touring China's giant battery factories Brinsden was convinced that the demand for lithium would be huge and that China would take the lead in electric vehicles. He went everywhere: 'from factories with lots of people and relatively little automation to those as big as three football fields with no people and all robots', he told me. While Tesla and Elon Musk got all the attention, the real growth was going to come from China, where the government had launched generous subsidies for EV purchasers. In 2015 China launched an ambitious plan to dominate future technologies, called 'Made in China 2025', which was a significant government-led effort to decouple China from global supply chains and support indigenous innovation. One of the key sectors was electric vehicles. That led to a rapid expansion in Chinese battery capacity. Over seventy percent of new battery cell capacity was being built in China.[9] 'It is not unusual to go to a battery factory and to be standing in the first of maybe one or two battery production lines but in the meantime, they have already installed sheds three through to eight and they are backfilling equipment and starting commissioning some of them,' he said.[10] Brinsden believed China would 'surprise the world' with the battery revolution. If you were away from China for as little as three months you would miss big changes in the industry, he said. Biddle shared his thoughts: 'We realised how much they were gearing up and in Australia we had no real knowledge of what was going on,' he recalled. 'I went to a couple of battery plants that were almost empty but they were gearing up to build batteries. These factories were enormous. That's what blew me away: the scale to which the Chinese were gearing up for an electrical future was mind-boggling. Ken was straight onto that; he grasped that really quickly.' Shortly after taking over Brinsden signed an agreement with a Chinese company, General Lithium, to buy unprocessed pure rock out of the ground from the mine – which would be placed on a ship to China. It contained 1.5% lithium – meaning Brinsden was just shipping basic rock to

China. But it would bring in the money while the company built its processing plant.

It was a race against time as other lithium mines were opening all around him in Western Australia. Altura, a former coal miner, was developing a lithium mine right next door. 'We can stand on the top of the hill and wave to them,' Brinsden said. In September Tesla's Elon Musk announced that its first mass-market electric car would be launched in 2016, the Model 3. Three months later analysts at Goldman Sachs called lithium the 'new gasoline'.[11] Lithium prices started to rise.

Brinsden quickly started to do deals, agreeing to supply lithium in return for finance to complete the mine, an industry practice known as an 'offtake'. It was good timing as prices for lithium carbonate and lithium hydroxide surged by double-digit percentages to near $10,000 a tonne. In early 2017 he signed a deal with Wang Xiaoshen, which would provide him with over half of the mine's annual output for ten years in return for $20 million. It was an agreement that gave Ganfeng a huge amount of lithium-bearing rock for relatively little investment. The same year China's car company Great Wall Motor, which had launched its first EV in 2017, bought a 3.5% stake in the company. 'Clearly there has been a structural shift in demand,' he told me, after raising A$100 million from retail and institutional investors.

In October of that year Pilbara delivered its first shipment of 8,800 tonnes of spodumene from Port Hedland to north Asia – under four years since the first hole was drilled. By 2018 the company was valued at $2 billion on the stock market, making millionaires out of its founders.

Thanks to companies like Brinsden's and investment from China, Australia became the world's largest lithium miner. Australian supply of lithium almost trebled between 2016 and 2020. The number of lithium mines went from one – the Greenbushes mine in Western Australia – to six. It was too much too quickly, and China was flooded by Australian lithium rock. By 2019 prices for lithium had started to collapse. Brinsden had expected the cycle to turn at some point – as it had in iron ore – but not that quickly. When I met him at my office in London at the beginning of 2020, he looked subdued. Not only had he contended with low prices but a contractor had also murdered a colleague on the mine site in 2019, forcing Pilbara Minerals to halt production. The mine was only operating at fifty to sixty percent of its capacity due to the weak market and twenty percent of the staff had been cut. 'China's caught a cold. So Western Australia's spodumene has got the flu, but it does not mean that the story is over,' Brinsden told a conference a few months earlier.[12] Adding to the woe, a few months later the Covid-19 pandemic hit. By October, the neighbouring lithium miner Altura went bankrupt. And prices for Australian spodumene continued to grind lower, reaching $400 a tonne, down from $900 a tonne in 2018. 'It's tough times for the industry, it's therefore unsustainable,' Brinsden said. 'Supply will not increase at today's prices; in fact the opposite, it will shrink.'

Brinsden turned out to be right. By the beginning of 2021, lithium prices had begun to rise in China and Brinsden bought the bankrupt Altura mine, then raised A$240 million from investors, including Australia's Super-Fund pension fund. By the summer of that year, as the world recovered from the Covid-19 pandemic, shares in Pilbara Minerals had risen by over four hundred percent, making it one of Australia's best-performing stocks.

Yet while the rise of the Australia–China lithium nexus had challenged the traditional lithium industry and changed geopolitics, it had also brought larger problems for global carmakers. As Ganfeng and others hoovered up lithium rock from Australia, the environmental toll of the process soon became apparent. Digging up lithium rock in Australia and roasting it in China had a carbon footprint as much as six to seven times higher than lithium from

Chile or Argentina, according to Benchmark Mineral Intelligence. That added to the carbon footprint of electric vehicles, just as sales were taking off. Digging up lithium from Australia involved conventional mining and crushing using fossil-fuel-powered vehicles. The lithium was then put on a ship run on diesel and sent to China, where the rock was processed using coal or natural gas and leached using acids. The Chinese lithium rush had shifted the centre of gravity of the industry, but also increased the carbon emissions. Making lithium in China was the most carbon dioxide-intense process in the world. While an electric vehicle would still significantly reduce the amount of carbon dioxide emissions over its lifetime compared to a petrol car (lithium producer Albemarle estimated that for every one kilogram of CO_2 emitted in its production of lithium, it enabled a greater than fifty times reduction in greenhouse gases due to its use in a battery), it was still bad news for a new industry that wanted to rapidly scale up production. If lithium hydroxide, the type used in Tesla's vehicles, were to keep being produced in China, it would be the equivalent to bringing on new carbon dioxide emissions equivalent to those of the entire country of Jamaica, according to one study.[13]

Yet by the time Tesla and other electric carmakers started to pay attention to where the lithium was coming from, it was too late. The Chinese-controlled supply chain worked smoothly and was cheap and efficient. In 2019 BMW signed an agreement to buy lithium from Ganfeng for five years, in a deal worth 540 million euros. 'This has caught some EV manufacturers off guard,' Alex Grant, a lithium expert, said. 'Their end customers are paying a premium for decarbonized cars, but EV manufacturers typically have minimal insight into the way the chemicals used in their batteries are made, with minimal resolution of the CO_2 emissions tied to extraction and processing.' Grant called it a game of 'carbon whack-a-mole' where we eliminate the CO_2 emissions from burning petrol, but substitute them for emissions elsewhere.

'What's the point of electrifying Norway's vehicle fleet if we have to emit hundreds of millions of tonnes of CO_2 in China? The atmosphere's greenhouse effect doesn't care where the CO_2 molecules came from.'

There was another problem too: just as Australia was making money selling lithium to China, political relations started to sour. Over the past few years, the Australian government had hardened its political position towards Beijing. Australia had quietly shifted from a policy of strategic cooperation to one of competition, as Geoff Raby, a former ambassador to Beijing, put it. 'The emerging dominant view of China was that it was seeking to overturn the US-led order in the region,' he wrote.[14] In 2018 Australia announced it would ban the Chinese telecommunications company Huawei from all 5G networks in the country, before any other Western country. The relationship with China reached a nadir in early 2020 after Australia's prime minister Scott Morrison called for an investigation over the origins of the Covid-19 pandemic in the Chinese city of Wuhan. China immediately reacted by suspending imports of various Australian products, including copper and coal. But they left iron ore and lithium alone. It was a reflection of China's dependence on Australia: the Chinese economy would collapse without supplies of iron ore from Australia for its enormous steel industry, and China needed the country's lithium to become a leader in electric vehicle technology. China produced over one billion tonnes of steel in 2010 – making it the world's biggest producer by a large margin. And likewise, China was the world's largest EV market. China's leaders had long fretted about their reliance on imported oil and gas since it exposed Beijing to sea lanes patrolled by the US, but it was a dependence that also extended to most mined metals and minerals too.

Brinsden was worried by the deterioration in the relationship. 'It's a cause for concern,' he told me. 'We've just asked for cool heads to prevail, to keep the megaphone out of diplomacy and to make sure to the extent possible that it's constructive.' The Australian Foreign Investment Review Board had already started to get stricter about Chinese investment in lithium and other battery minerals. In 2020 it blocked a Chinese lithium company, Yibin Tianyi Lithium (a company backed by China's largest battery maker CATL), from buying a stake in an Australian-listed company that owned a lithium project in the Democratic Republic of the Congo. Wang's company, Ganfeng, had also agreed to buy lithium from the project, which was operated by a company called AVZ Minerals. As a result of Australia's stricter stance, Chinese investment into Australia plummeted by sixty-one percent in 2020.

Australia began to court Europe and the US, in the hopes of becoming a key supplier to their emerging clean energy and electric vehicle supply chains. In late 2019 I met Bill Johnston, Western Australia's Minister for Mines, as he toured Europe trying to drum up investment in battery minerals. 'There are many opportunities in Western Australia, they're not all taken,' he told me at the Australian embassy. 'While we welcome Chinese investment we're very keen to welcome alternative investment. There's almost nothing that goes into a battery that we don't produce in Western Australia. You can look at the periodic table and I've probably got a world-class project for you.'[15]

But China had already started to look elsewhere. Wang had no intention to just rely on Australia, and he realised that the high carbon intensity of Australian lithium mining would one day be unpalatable to European and American carmakers. In 2017 the American lithium expert Lowry had introduced him to a Canadian junior miner, Lithium Americas, which was working on a project in Argentina with SQM, which planned to evaporate lithium using sunlight from lithium-rich brine. Ganfeng agreed to pay $174 million towards the project, in return for seventy percent of the future supplies of lithium, and took a twenty percent stake in the Canadian-listed shares of Lithium Americas. The following year Ganfeng bought SQM's fifty percent stake in the Argentinian project for $88 million. It was a scenario unthinkable a decade earlier, when SQM supplied lithium to Ganfeng. In early 2020, as relations between Australia and China deteriorated, Wang increased his holdings of the Cauchari-Olaroz lithium project in Argentina to fifty-one percent. Situated 4,000 kilometres above sea level in one of Argentina's poorest areas, in the far north-west Jujuy province, the project would be Ganfeng's first fully owned and operated foreign project. In the same area a solar power plant was being built by a Chinese construction company and backed by the Export-Import Bank of China. With 1.2 million solar panels it was among the largest solar plants in South America.

When I met Wang in his bare office near the Science & Technology Museum in Shanghai on a sunny day in early 2019, he had just come back from meeting with Argentina's president Mauricio Macri. Macri had welcomed the idea of using renewable energy for the lithium plant, Wang told me. Ganfeng had secured the rights to develop another project in Argentina a few years back too, Wang said, called Mariana. Because it was even more remote, the company was considering using its own batteries and building a solar plant. After we chatted, we left the office and walked in the sunlight to a nearby shopping centre to have lunch in a crowded café. The shopping centre had fifty Tesla chargers, Wang told me. Tesla's construction of a Gigafactory in Shanghai had been a boost to Ganfeng's business, he said, as the Shanghai government preferred Tesla to choose local suppliers. A year earlier Ganfeng had signed a three-year deal to supply lithium to Tesla. Ganfeng was one of the few lithium companies that was expanding capacity and growing production. 'They were ahead of the curve, they were true believers early on and pursued this vision,' John Kanellitsas, vice chairman of Lithium Americas, told me. 'Only now is it becoming real in the eyes of the West.' Wang would soon send a team of Ganfeng experts to live high up in the Andes and build the processing facilities for the new Cauchari project, which he hoped would help Ganfeng supply the coming wave of EVs. 'If you look at the long term,' Wang told me, 'definitely this demand is coming.'

China had arrived in Latin America.

6

Chile's Buried Treasure

In early 2018, I got the chance to meet Chile's Lithium King. In the dying light of a Santiago dusk I took a taxi through the suburbs, past the polo grounds where horses were being led on their evening stroll. The taxi stopped opposite the grounds in front of a smart cream-coloured block of flats, with its own concierge. I entered and was ushered discreetly upstairs. This was the home of Julio Ponce Lerou, a Chilean forestry engineer who had married one of the daughters of Chilean dictator Augusto Pinochet and during his rule took control of one of the world's largest producers of potash and lithium. In the process he became a billionaire and one of the richest men in Chile – worth some $4.7 billion at the time of our meeting, according to *Forbes*, controlling the world's lowest-cost producer. From this flat, however, Ponce Lerou was trying to fight a rearguard battle against a takeover from China's Tianqi Lithium, in the face of an enemy he could have never imagined: a short, tough woman named Vivian Wu. As I entered his apartment, the global race to control the supply of lithium was only just beginning. And it had all the hallmarks of a classic gold rush.

Trim and athletic with closely cropped grey hair, Ponce Lerou exuded the confident energy of a wily operator. He had helped build Sociedad Química y Minera de Chile (SQM) in the desert of northern Chile into one of the world's largest suppliers of lithium and fertilisers from a position of near bankruptcy, and had dedicated his life to the company. 'Julio is an atheist, but he believes in SQM,' one of his confidants told me. But along the way, he had also become the figurehead for the cosy corruption and elitism of Chilean politics, highlighted by a series of scandals that broke out in 2015, which showed that some of Chile's largest companies were illegally financing politicians across the board. 'They should nationalise SQM and take him out,' one copper mining executive told me. Ponce Lerou was the person everyone loved to hate.

Between 2008 and 2015 SQM had made almost $15 million in 'improper payments' to a host of Chilean politicians and political candidates, according to an investigation by the US Securities and Exchange Commission. But it was his twenty-two-year marriage to Verónica Pinochet that to many in Chile was Ponce Lerou's 'original sin'. In almost every media article he was still mentioned as Pinochet's son-in-law in a country that has a complicated relationship with the man who ruled Chile brutally for almost two decades before being deposed in 1990. More than 3,000 people are estimated to have been killed during Pinochet's rule, which began with a military coup in 1973. It was under Pinochet that SQM had been privatised, due to the influence of a group of Chilean economists nicknamed 'the Chicago boys', after the Chicago school of free market economics where many of them had studied. Like the Russian oligarchs who grew rich with the privatisations of the 1990s after the dissolution of the Soviet Union, Ponce Lerou had also ended up with control by buying up shares that had originally belonged to the workers. It seemed bitterly unfair to people that one of the main people set to cash in on the lithium boom was Ponce Lerou, the *yerno* (son-in-law).

But he had no intention of stepping aside. It was the Chinese that were now giving him trouble. China's biggest lithium producer, Tianqi Lithium, wanted to take a large stake in SQM – and possibly control, giving China a dominant position in the global lithium supply chain. For years Chile had been happy to sell copper to China, but this was different. This was the beginning of a technological revolution, and the battle was on for the commanding heights. As he escorted me out and rushed across the road to his car to drive to the airport and catch a flight to San Francisco, I realised that this is where the clean energy age begins. Those who control the resources we will need in vast quantities to avoid the catastrophic impact of climate change will be the new Rockefellers. A new strategic game had opened up. And China, which had already spread its tendrils across Latin America through investments and loans, had a head start. Ponce Lerou had a formidable rival.

*

A 600-mile-long strip of land between the Pacific Ocean and the snow-capped Andes, northern Chile's Atacama is the world's driest non-polar desert. Much of it is hostile to human life, and rainfall has yet to be recorded in some areas. For hundreds of years men have flocked into the desert to search for minerals, ever since the Spanish arrived in Chile in the sixteenth century. First it was the lust for gold that drove them, just as it drove the conquistadors across Latin America. Then came nitrates, a critical fertiliser that fed the world – making the Atacama so valuable it caused a war between Chile, Bolivia and Peru. During Pinochet's reign, the desolate desert was where he sent his political enemies to be executed, their bodies placed in mass graves. Now the Atacama is set to play a key role in the clean energy economy. The desert has the highest concentration of solar radiation in the world: providing ever cheaper solar power to Chile. It's also the world's biggest source of copper, from mines dotted across the

Atacama, a metal which will be crucial for wiring up electric cars and the required charging infrastructure – the average electric car contains eighty-three kilograms of copper, over three times the amount in a petrol-powered one. And beneath the desert floor lie some of the world's largest reserves of lithium. The lithium is concentrated in brine that was washed down from the surrounding mountains millions of years ago. 'We are poised to become a central actor in the creation of the new electrical world economy,' Óscar Landerretche, the former chairman of Codelco, the world's largest copper producer, owned by the Chilean state, told me in the company's grand dark-panelled wooden offices in central Santiago. The Atacama is the beginning of a global supply chain that enables all our digital lives – from iPhones to iPads and laptops. It is a reminder that for all our ability to live on our smartphones and store our data in the cloud we have still not moved beyond digging up finite minerals from the earth to meet our needs. As Max Planck once observed: 'Mining is not everything, but without mining everything is nothing.'[1]

In early 2016 I was picked up in the dusty mining town of Calama, where copper miners once spent the fortunes they made from feeding China's insatiable demand for the metal in the mid-2000s. They used to fill the bars, the *choperias*, and the more seedy strip clubs in town, spending their enormous bonuses. Copper money had funded a new glitzy shopping centre and a casino. But in 2016 Calama had the air of spiritual emptiness that so often accompanies boom towns that have fallen on harder times. Copper prices had plummeted, and workers had lost their jobs. These pressures had culminated in the longest mining strike in Chile since Pinochet came to power in 1973. I accompanied one retired copper miner to a hill overlooking the town where there was a huge statue of Christ donated by the Argentinians. It was here that people came to paste notices thanking the 'Christ of the desert' for wishes that came true in their lives.

Instead of copper, workers now flocked into the desert to mine for lithium. Every week about a thousand workers travelled by bus to a site in the north of the desert to work seven-day shifts on the *salar*, the thousands of miles of salt flats that shine white in the glare of the sun. Most came from the nearby regions of Tarapacá and Antofagasta. They slept in corrugated huts, in a small compound that had its own football field and an outdoor stage for film nights. Twenty-four hours a day, every day of the year, a salty solution rich in lithium was pumped from deep beneath the desert into evaporation pools. The concentrated bright green brine that was produced was then driven west in small trucks to processing plants on Chile's coast. There, it was refined into a powder and placed in large white bags before being sent around the world. Much of it travelled across the Pacific to China, where it was used to create rechargeable batteries that power hundreds of millions of smartphones, digital cameras, laptops and electric cars, as well as large energy storage facilities to enable the transition to renewable energy.

Drenched all year in a piercing sunlight, solar radiation in the Atacama makes it perfect for extracting lithium and also for solar power. Dotted clear blue and green across the Atacama, the man-made pools filled with brine looked like the scheme of some outlandish billionaire to build swimming pools in the desert. While the brine-extraction process is slow, it is generally a cheaper process than hard-rock operations, since the lithium is already isolated within the brine, and the sun does most of the work. Lithium is also produced as a by-product of the production of potassium chloride, SQM's core product, which is used to make fertiliser. SQM, or Soquimich as it's known in Chile, has built over forty-four square kilometres of evaporation ponds, making them the largest artificial ponds in the world. Interspersed with mounds of salt that are produced from the evaporation process, the ponds sit right in the middle of the desert in an ecologically sensitive area, where flamingos feed off the tiny crustaceans in the surrounding lagoons.

Standing beside the pools dressed in a shirt and hard hat I felt the heat burning my skin. It also hurt my eyes to look out over the pools of bright green water. Alejandro Bucher, a young energetic employee from SQM, let me touch the end result: the lithium-rich bright green brine, which felt itchy against the skin. Standing on one of the mounds of salt, which have been made into viewing platforms, Bucher listed the facilities the company had to provide for the workers in such a remote location, from a specially built hospital to sunscreen. The crusted salt pans of the desert isolated the plant like an inland sea. No foundations could be built, so telephone lines ran on top of man-made mounds, as did huts and buildings. The executives stayed in a smarter hut on top of the salt, a cluster of well-cleaned rooms with a private chef. This was where Ponce Lerou would stay when he visited frequently while he was SQM's chairman. He had taken an obsessive interest in the operations – and knew the intricate details of the global fertiliser market. But when I visited in 2016, workers' morale had been hit by a corruption scandal, which was in the Chilean press on an almost daily basis. Patricio Contesse, Ponce Lerou's old friend and the CEO of SQM, had been forced to step down. Contesse had allegedly used a total of $14.75 million from his personal 'discretionary budget' to pay off Chilean politicians, mostly using fake invoices. Many of these payments were made to foundations run by politicians of all colours, including the former Economy Minister Pablo Longueira. Longueira received over $630,000 from SQM, including $16,000 paid into his daughter's foundation, according to an investigation by the US Securities and Exchange Commission. Other officials involved had been placed under house arrest by Chilean investigators.

Many Chileans were calling for SQM to be banished from the Atacama Desert completely. The man threatening to take SQM's lease away was Eduardo Bitrán, a bearded economist who grew up in the small town of Ovalle in northern Chile. He trained as an industrial civil engineer at the University of Chile during Pinochet's dictatorship and then later received a PhD from Boston University. Growing up surrounded by poverty, Bitrán decided to go into academia instead of business, with the hope of joining the government when Chile eventually returned to a democracy. His ambition was to make Chile a developed, middle-class country. Pointed in his opinions, which he often voiced on Twitter, Bitrán had a hatred of Ponce Lerou, and everything he stood for. Ponce Lerou 'represents a very unhappy time in our history, where a few took advantage of the pain and horror of many', he said.[2] As head of Corfo, the country's economic development agency, Bitrán was in charge of renegotiating SQM's lease to extract lithium from the Atacama Desert (its major competitor Albemarle had already done so in 2016). When I visited him in his office, he could barely conceal his hatred for the company and Ponce. SQM had not complied with the terms of its rental contract to extract lithium, he said. 'I think the most difficult problem now...is the big problem with SQM, in terms of the fact that they have been playing complex games in the Chilean political system,' Bitrán told me. 'We, as the owner of the *salar*, want to make an alliance for the exploitation [of the Atacama] with companies that behave according to the rules and according to international standards in terms of corporate governance, in terms of compliance.' In 1979, lithium was labelled a 'strategic mineral' in Chile, in recognition of the fact that an isotope of the metal is used in nuclear fusion. The label stands to this day, despite the fact that the nuclear industry now uses an inconsequential amount of lithium. That means the government retains the rights to the lithium in the *salar*, which it leases out, in contrast to copper, where companies can obtain concessions that give them direct rights over the minerals in the ground. The nuclear commission also restricts the total amount of lithium that companies can produce through a quota; SQM can't extract more lithium from the *salar* than it agreed in the terms of its lease with the government in 1993. When I mentioned Bitrán's complaints to Bucher, however, he brushed off these concerns. 'I don't think that kind of Venezuela-style approach is possible here in Chile,' he said as we ate steaks in the executive hut perched on the salt flats. I would be contrary to Chile's reputation as the best-run mining country in the world. Outside, the surrounding volcano-studded mountains grew dark and the stars appeared, piercingly clear above the desert. The night sky in the Atacama is the world's best place for astronomy. A cluster of powerful telescopes dot northern Chile from where scientists peer deep into our solar system, looking for stars and signs of habitable planets.

Julio Ponce Lerou's early life had been the opposite of Bitrán's. He was born into a middle-class family in La Calera in central Chile in November 1945, the son of a doctor and nurse of French descent. During the holidays the family travelled to the resort of Maitencillo on the coast of Chile, and it was there that Ponce Lerou met Verónica Pinochet Hiriart, when she was fifteen. The couple would later marry in 1969, when Ponce Lerou was twenty-four. After boarding school in Santiago Ponce Lerou studied forestry at the University of Chile, where he met Daniel Contesse, the brother of Patricio, who would later go on to be chief executive of SQM. After university he worked at a sawmill in Ontario before returning to work for the paper and pulp company Inforsa, and CMPC, one of the country's largest paper companies. The election of socialist Salvador Allende as president in 1970 came as a shock to the young and ambitious businessman, as it did for many in Chile. The country suddenly lurched to the left. Cuba's leader Fidel Castro paid a long and widely publicised visit to the country in the following year, stepping off a white Soviet jet aeroplane before being received with military honours and a twenty-one-gun salute. Copper miners were urged to 'defend the revolution through more production' and mines were nationalised by the agreement of Congress.[3] Ponce Lerou decided the 'Chilean Road to Socialism' was not for him and left the country. He moved to Panama to take over a sawmill in 1979. It was from there that he and Pinochet's daughter watched the US-backed military coup that deposed Allende a year later, leaving him dead in the presidential palace in downtown Santiago. General Pinochet called Ponce back to Chile to become an executive director at the state-owned National Forestry Corporation a year later, a turning point in his career. Ponce quickly rose up the ranks of the new corporate elite in the country, notching up a series of financial successes as Pinochet tightened controls over the media and the military regime hunted down its enemies. At the age of thirty-two he became head of another state-owned cellulose company that was privatised. In 1979, as Pinochet pushed through his programme of economic reform, driven by the 'Chicago boys', Ponce became head of Corfo, the agency tasked with reversing the nationalisations that had taken place under Allende. As Corfo's representative he also became the de facto head of some of Chile's largest companies, such as the Telephone Company of Chile. It was at this time that he came across SQM, a once grand but failing company founded in 1968 and nationalised by Allende in 1971 that produced potassium-based fertilisers.

At the company's Nitrate Building in Santiago, waiters dressed in black with white gloves still served food from silver platters on the top floor, in an office decked with fine art. In its City of London offices gin and tonics were still standard at noon. SQM had a total of 10,000 workers but wasn't making a dollar of profit. In 1981 the company's board and executives were changed and Ponce joined as chairman. 'If that decision wasn't taken then SQM would not exist today,' Ponce said in the company's official history.[4] His friend Patricio Contesse joined as chief executive, and Eugenio Ponce, Ponce's brother, joined as commercial vice president. They remained in the company for the next thirty years.

The following year the company made an accumulated loss of $44 million due to weak demand on international markets and subsidised prices for its fertilisers in Chile.[5] A year later the privatisation process began. The slate was set for a recovery and the new board engaged in a wholesale restructuring, selling company assets and cutting costs. The number of workers was cut down to 4,000 and austerity imposed on the company. It started to make a profit, and listed some of its shares on the local stock exchange. Ponce Lerou resigned from Corfo in July 1983 following allegations of corruption, and took a break from state business, instead buying up cattle to try and start an agricultural company. But that same month he bought his first shares in SQM, at rock bottom prices. He would be back. SQM was at the beginning of a five-year privatisation process, which would eventually lead to Ponce Lerou holding majority control of the company.

The privatisation was initially designed to give workers shares in their companies under the slogan 'popular capitalism', with a promise to make workers owners of their companies. At the time the labour law stated that all companies had to give ten percent of the profit every year to the workers, in proportion to the salaries they received. At SQM workers were awarded part of the company's profits in shares. They were then encouraged to put their shares into a new investment company, called Pampa Calichera, which would hold them until they retired or left the company when they would get them back. The state, through Corfo, sold its 48.56% holding in SQM for $140 million in October 1986 (a fraction of the value of the company today but considerably more than the $80 million value of the company in 1983) and by March 1988 SQM was completely privatised. Pampa Calichera ended up becoming the largest shareholder after continued purchases funded by bank loans backed by the workers' shares. It was through this and other holding companies, known as the *cascadas* or 'waterfalls', that Ponce Lerou built up his control of SQM, ending up with a thirty percent stake in SQM in the 1990s. Many workers later said they were pressured to sell their shares. In 1989, workers were even courted by models in miniskirts to encourage them to sell their shares, according to one former worker.[6] Eduardo Bobenrieth, who had set up Pampa Calichera in 1986, told me that all of the workers had been cheated or pressured out of their shares.

The privatisations under Pinochet remain controversial, even though they did lead to an increased corporate performance in many cases, and especially in the case of SQM.[7] Pinochet's 'Chicago boys' could point to improvements in the Chilean economy in the 1980s, as well as to similar moves to privatise state assets in Margaret Thatcher's Britain and in the US. Ponce argues that he rescued SQM from failure and that its shares were cheap because the company wasn't performing well. But the privatisations have left a lingering sense of suspicion in Chile along with accusations that the state sold off its best assets too cheaply. A report from the comptroller general's office that came out after the fall of Pinochet in 1990 said the state had sold parts of SQM for less than a third of their fair market values at the time.[8]

Ponce Lerou's history was anathema to Bitrán, who was friends with Ricardo Lagos, a key opponent of Pinochet who became president in 2000. Bitrán believed that Chile should be ruled by law and that the government and the people should benefit more from its lithium resources. He wanted Chile to become a renewable energy superpower, extracting lithium but also using its plentiful supplies of solar power to build batteries in the desert. 'In early 2016 I started to realise that Chile would be facing an opportunity that we never thought about,' he recalled. 'We had an opportunity to transform the Chilean economy.' In May 2014 Bitrán had filed a legal arbitration to terminate SQM's contract to extract lithium from the Atacama. Bitrán accused the company of breaching the terms of its lease agreements, under which it has the right to extract lithium. After years of legal threats from both sides, in January 2018 he signed an agreement with the company, which came with multiple governance, social and environmental conditions, including that Ponce give up his right to sit on the board. The agreement also significantly increased the royalties SQM had to pay the government on its lithium sales, especially if lithium prices went even higher. SQM also had to contribute between $10 to $15 million for neighbouring communities in the Atacama, as well as agree to sell lithium at a preferential price to companies who wanted to make batteries or battery parts in Chile. Bitrán didn't want Chile just to become a producer of raw materials, but also to capture some of the additional value of the electric car supply chain that was mostly going to China. It was a bold government-led attempt not only to gain greater revenue for the state from the country's raw materials but also to diversify the country's economy.

But there was one problem: Ponce Lerou did not quit the scene. A few months after the election SQM announced that Ponce and his brother Eugenio had rejoined the company as 'strategic advisors' – and not as members of the board. It was a provocation, carried out against everything Bitrán had tried to achieve. It took SQM back to the heart of the Pinochet era, as if nothing had changed. The decision prompted instant criticism from politicians in Chile. Faced with a significant media backlash SQM backtracked and rescinded the appointments. But Ponce Lerou remained the largest investor in SQM. In late 2018 he travelled to the New York Stock Exchange to celebrate the twenty-fifth anniversary of the company's listing, standing behind the chief executive as he rang the opening bell. SQM was in many ways still a relic of his era: its board included Hernán Büchi, a free market ideologist who had pioneered Pinochet's economic reforms.

Ponce Lerou would soon face his own reckoning, however, and this time from across the Pacific. Just as Ponce Lerou had managed to control a formerly state-owned asset in Chile, in China's mountainous Western Sichuan province, another man, Jiang Weiping, had managed to take hold of a state-owned lithium company in the small county of Shehong, making his fortune. In 2004 Jiang agreed to buy the Shehong lithium company from the local government for a price of over 11 million yuan ($1.7 million at today's exchange rate). While Jiang immediately took control of the local government's lithium assets he didn't pay up the full amount until over two years later.[9] Also, the local government did not seek confirmation for the privatisation from the provincial authorities until a full four years later, in contravention of the rules for buying and selling state-owned assets. The process had taken place under what Chinese call 'act first and report later,' one reporter said.[10] Founded in 1995, the Shehong lithium company had been making losses so the local government was keen to get rid of it. The global battery market was small in the early 2000s, before the widespread adoption of mobile phones, and lithium prices were also low. Jiang's timing was impeccable, however. The first iPhone came out in 2007, and smartphones became ubiquitous a few years later. After Tianqi Lithium listed on the Shenzhen Stock Exchange, Jiang became a billionaire in early 2015, reaching the *Forbes* rich list for the first time.

Jiang was born in 1955, six years after Mao Zedong and the Chinese Communist Party came to power following a bitter civil war with the Nationalists led by Chiang Kai-shek. A member of the Bai ethnic group, Jiang went to school in Chengdu and at the age of nineteen was sent to work at a commune during Mao's tumultuous Cultural Revolution, as part of the wave of 'educated youth' who were sent down to the countryside. In 1978 after the end of that period of upheaval he entered the Chengdu College of Agricultural Machinery, graduating with a degree in engineering in 1982, just as a period of greater liberalisation and economic reform was underway. He started work as a technician at the Chengdu Machinery Factory, moving on to another machinery company a few years later before starting his own business in 1997. But it was the deal to acquire the lithium processing assets from the local government that made his career, making him one of the richest men in Sichuan province. For his services Jiang served on the political committee of Shehong county. Like many business-men in China, he wisely kept his government and Communist Party connections.

Jiang also heeded the Communist Party's call to 'go global' and secure overseas natural resources. Along with Vivian Wu, the young Tianqi president, Jiang got his second big break in 2012, when he won a bidding battle against US lithium pro-ducer Rockwood – now Albemarle – for control of Australia's largest lithium mine, Greenbushes, paying $646 million in a deal backed by China's sovereign wealth fund and largest state-owned policy lender China Development Bank. The mine had been in operation since the days of Australia's gold rush in 1888, producing first tin and then tantalum, but it had been bought by a US investment fund following bankruptcy in 2007. Tianqi had been one of its main customers for lithium, buying forty percent of its supply to process in China. So Jiang was concerned when his much larger US rival Rockwood made a 2013 bid for the company. Jiang worried that the US company would secure supplies of lithium just as the battery boom was set to accelerate. Tianqi hired Andrew Low, an Australian with a keen interest in China who had been educated at Tsinghua and Harvard Business School, to help with its bid and smooth its political approval in Australia. Low jumped on a plane to Chengdu to visit Tianqi. Tianqi started to secretly buy up shares of the Canadian-listed company that owned the Greenbushes mine, Talison Lithium, on the open market. Then, once it had breached the ten percent threshold for disclosure, Jiang offered a price for Talison that was a fifteen percent premium to Rockwood's offer, backed by financing from China Development Bank. Most Western companies cannot get a vast state-owned bank such as CDB to provide credit for a deal, the equivalent of a sovereign guarantee. The investors who owned Talison were only too happy to accept and the deal was waved through by Australian regulators. The Tianqi team celebrated in Chengdu with a ten-course banquet washed down with the Chinese liquor Maotai.

*

It was the first round in the global race to secure lithium supplies and Tianqi had won before anyone could pay attention.

Having secured the world's largest lithium mine in Australia, Tianqi started to look at Chile. It took its time with SQM, first buying a two percent stake in the company in late 2016. Tianqi approached Ponce Lerou and offered to buy his stake in the company, given the pressure surrounding the corruption scandal in Chile. The talks quickly fell apart, however. So Tianqi waited for another opportunity. One duly arose one year later when Canadian fertiliser company Potash Corp, a large shareholder in SQM, made a bid to acquire its rival fertiliser maker, Agrium. Potash was forced by Indian and Chinese regulators to sell its shares in SQM in order for the deal to go ahead. In late 2017, Tianqi presented an offer to Bitrán for Potash's stake, the day after the election of Sebastián Piñera as president of Chile. Bitrán was not in favour of a bid by Tianqi for SQM, fearing that the Chinese company would eventually seek control. In one of his last acts of business under outgoing Chilean president Michelle Bachelet, Bitrán filed a complaint with Chile's antitrust regulator seeking to block the sale of the shares to Tianqi, on the grounds that they were a major competitor. But Piñera had campaigned on a pledge to revive foreign investment in the mining sector, and following his election

Tianqi received a signal that its bid would be welcomed. In May 2018 Tianqi formally offered $4 billion for twenty-four percent of SQM, a huge amount for a minority stake in a company. Tianqi would have three seats on the board of SQM, giving the company an inside view on Chile's biggest lithium producer.

Ponce Lerou was not happy. For years he had controlled the board, but now thanks to both Bitrán and Tianqi, he was set to lose his influence. From his perch in Santiago he mustered all of his government connections to mount a counter-attack, accusing Tianqi of having access to SQM's commercial secrets if the deal went through. He filed his own legal action with Chile's Constitutional Court seeking to block the transaction. He had a point: Tianqi was a serious competitor to SQM, via its ownership of the Greenbushes lithium mine in Australia. It was an argument that fell on deaf ears in Piñera's Chile, however. Bankers who worked for Tianqi were equally enraged. 'This guy courted Tianqi a few years ago when he was in financial difficulty,' one banker told me. Unlike his previous battles, this time Ponce was fighting the Chinese government. The Chinese approach to a key Chilean asset briefly put Bitrán on Ponce's side, while other experts such as Chilean lawyer Alonso Barros welcomed Tianqi's dilution of Ponce's control over the industry. In April of 2018, China issued a stark warning to Piñera when in an interview with local media Xu Bu, the newly arrived ambassador, told local newspaper *La Tercera* that the opposition to the stake sale could 'leave negative influences on the development of economic and commercial relations between both countries'.[11] Soon enough, Chile's court waved the deal through, just as regulators had done in Australia a few years earlier. From then on, SQM sought and largely managed to separate its corporate image from Ponce.

As Chile looked set to benefit from the lithium boom it found itself ceding influence over its largest lithium company to China. Its crown jewels had been sold off by Pinochet, and then once again by Piñera's government. It was a disappointment to Bitrán. 'It was obvious the Chinese wanted to take it over, to monopolise lithium and cobalt in order to have the domination of electro-mobility,' he said. Chileans were left with Bitrán's promises that Chile would use its lithium to become a leader in manufacturing and renewable energy, but the country had received only a few bids from battery makers. No large Chinese battery maker had offered its services. It made no sense for Chinese companies to shift production across the Pacific – all they needed were Chile's raw materials, just as the world had needed Chile's supplies of nitrates to make fertiliser in the nineteenth century. In the race for control of the battery supply chain, China had found that it could easily buy its way into dominance through the open market. All it needed was cash. With foreign mineral resources secured, China was now in a powerful position to expand its EV market rapidly. Chile risked becoming just another node in China's global supply chain intended to boost its dominance in electric cars.

By the time of Tianqi's deal, however, Western car companies had started to realise they needed large amounts of raw materials for their batteries. This awakening, however, began not with lithium but another battery metal: cobalt.

7
The Cobalt Problem

'The electronics industry is much more [like] casual sex with contractive negotiations whereas the automotive is much more about marriage and long-term commitments – it's a massive cultural change; a change in culture and size.'

Cobalt trader

'Cobalt is still a gotcha ... realise that cobalt is cocaine right now and there's a new cartel forming.'

Former BYD America's vice president Michael Austin at
Benchmark Minerals Week 2018, Newport Beach, CA[1]

As they travelled to Germany in late 2017 the world's small coterie of cobalt traders thought they might do the deal of a lifetime. The German carmaker Volkswagen had invited them to its imposing headquarters in Wolfsburg, one of the largest car production sites in the world. Built in the 1930s the five-square-kilometre campus was dominated by the brown-red chimneys of VW's main factory – a symbol of the company's manufacturing prowess, which had propelled Germany into the world's pre-eminent producer of cars. Volkswagen produced one in eight of all passenger cars sold in the world. But the German stalwart had been badly burnt by revelations two years earlier that it had deliberately installed cheat devices on its diesel vehicles to mask the amount of toxic pollution they emitted. Now Volkswagen was attempting to leave that past behind by going all-out electric. VW's website captured this anxiety: 'Think of an environmental role model,' it asked. 'Did you think of us at Volkswagen? Probably not.' It had boldly pledged that half of all its car sales would be battery-powered by 2030. But to do that it needed a huge amount of cobalt, a dull grey metal that is mined mostly in the Democratic Republic of the Congo, one of the poorest countries in the world. The carmaker did not know much about the obscure metal, which had been an afterthought in global commodity markets for the last few decades. It hired a London-based consultant to teach company employees about the market through a series of PowerPoint presentations.

It turned out that only a small number of people traded cobalt or gave it much attention – and almost all of them had been invited to Wolfsburg. That was good, as VW wanted to buy a lot of the metal, and was banking on signing a quick deal. The carmaker was in a rush: Elon Musk's Tesla Motors was garnering a huge amount of publicity and looked as if it might actually succeed.

VW was not the only carmaker pledging to go electric. One of the traders had been spending half of his year on the road meeting European and American carmakers, all of whom were making bold pledges. Tesla planned to open factories in Shanghai and Berlin and begin producing its own battery cells. On a recent trip to China, an executive from one of the country's giant battery companies had written the amount of cobalt they needed on a napkin, and he had almost had a heart attack. For years, cobalt, due to its high melting point, had been used in so-called 'superalloys' to strengthen jet engine blades – a market with fairly steady and predictable demand. But this was something different. The executive had asked for as much as 20,000 tonnes of cobalt by 2020, more than the entire market for alloys at the time.

The group invited to VW also included Franck Schulders, the energetic French head of cobalt for the Switzerland-based mining giant Glencore, the world's largest producer of cobalt. Schulders, like most of Glencore's traders, was smart, quick-thinking and aggressive. Anyone in the global commodity markets knew not to mess with Glencore, a company that supplies and trades most of the raw materials we need for everyday life from its headquarters in the sleepy lakeside town of Baar, until 2021 overseen by the billionaire South African Ivan Glasenberg.

Volkswagen led the traders to five separate executive boxes in the giant VW Arena football stadium, where they were told to sit and wait while executives from the sustainability and commercial departments talked to each of them in turn. Four hours went by, and VW had not moved on from the first supplier, the Chinese cobalt producer Huayou, which a year earlier had been accused by Amnesty of buying cobalt from child miners. The traders quickly grew bored, walking down and sitting in the football seats by the pitch waving to each other. They begged to be able to go out to a pizza restaurant for lunch, which shocked the VW executives, who were used to dealing with steel companies who competed fiercely with each other and didn't eat together like old friends. This small group *were* the entire cobalt market. They were the new frontiers of electric car production and the old rules no longer applied.

In the afternoon they received a visit from the executives, including Bastian Brodesser, a VW loyalist 'who probably has VW cut into his arm', according to one of the traders. Brodesser, a young man with short cropped hair, told him they wanted a 'VW discount' for the cobalt they needed. 'It's ok for you to sell to Tesla but we want a discount because *we are Volkswagen*,' he said proudly, according to people present. They were told to spend the evening talking to their management and come back the following day with a generous offer. By the end of the day the traders communicated their frustration to VW. The VW executives relented and let the group go for dinner. They would meet again the next day. Schulders from Glencore was not happy about that – he needed to leave the next morning so requested his meeting at 8.30 a.m. The others also chimed in, saying they could not wait around. That evening the traders went out to a local pub in Wolfsburg, where they ordered $100 bottles of wine. 'We were shell shocked,' one trader recalled. 'It was like being kept in a prison camp, without knowing when we could escape. Seeing the VW sign still gives me the shivers.'

The carmaker had wanted to buy years of cobalt supply at fixed prices, something that the traders dismissed instantly. 'They know actually nothing about buying cobalt,' one said. 'What size the bags are, how we transport it. They can't even tell me which processors they want to send it to.' In addition, Volkswagen had wanted the metal to be sent direct to Wolfsburg, the factory door, but the mining companies wanted to be paid once it had left the Congo. The group left without agreeing to sell a single pound of cobalt.

Volkswagen had treated the traders and miners like they were makers of widgets. The carmaker was used to dealing with just-in-time suppliers who gave discounts to be a part of their giant global supply chain. They had assumed that the miners would come running, happy to be a part of the electric future. They were the largest carmaker in the world. But it had failed to appreciate how mining worked – that it costs more to dig up more metal since mines cannot be expanded in the same way a factory line can. More supply would cost more. To the small group of traders, they had the upper hand: they were the ones with the cobalt that VW needed to make every single electric car. 'It's not that we're not willing to sell but why should I sell at a discount? They just don't get it.' Within twenty-four hours, VW, the colossus of the car industry, had blown it.

The miners left Germany without any regret. A few weeks later during the annual London Metal Exchange (LME) week, the key industry gathering for the world's metal traders, the traders gathered at a private members' club off Berkeley Square, where cigar smoke was thick out on the balcony and champagne flowed freely inside. They had reason to be happy: the mining companies held all the cards. They were at the beginning of a long and complicated supply chain that carmakers were becoming wholly dependent on.

The meeting in Wolfsburg marked an awakening for Volkswagen. For Volkswagen and other carmakers, they would quickly discover the road to their electric car dreams was paved with corruption, child labour and geopolitics. As they rushed to go electric, car companies quickly found they had to outsource all of the key components of the electric car. They had to buy batteries from Asia, from giant companies like LG Chem, Panasonic, Samsung SDI and China's CATL, which in turn bought their battery raw materials from further down the chain. In 2017 Europe had no large-scale battery factories of its own, no lithium or cobalt mines, and only a little production of battery materials. The US was in a similar situation. Chinese car companies had also started securing the batteries and raw materials they needed, helped by China's huge presence at every stage in the supply chain. The miners were agnostic about who they sold to and only cared about the price. There were billions of dollars waiting on the table. Analysts predicted that if each of the one billion cars currently on the road were to be replaced with a Tesla Model X, demand for cobalt would equal fourteen million tonnes – twice the size of the current global reserves.

Over sixty percent of the world's cobalt came from the Congo, making it impossible to envision a rapid transition to electric cars without being completely dependent on the country, which was one of the poorest and most corrupt on the planet. A few months after the Wolfsburg meeting the *Economist* featured the Congo on its front cover with the headline: 'Heading back to Hell'.[2] In the DRC there was an unenviable choice facing car companies and battery makers. They could buy cobalt from the large industrial mining companies like Glencore. But the Swiss company was under greater scrutiny due to its connection to alleged corruption in the Congo, via an Israeli diamond businessman, Dan Gertler, who was a friend of the former president Joseph Kabila. (Gertler's background will be explained in greater detail in the next chapter.) Or they could buy from Huayou and other suppliers who sourced from thousands of individual miners in the Congo, who dug for cobalt by hand for $1 to $2 a day, without safety equipment and often alongside children. Buying cobalt through that route could expose the company to allegations of child labour, which could ruin the electric car's reputation. How willing would carmakers and battery companies be to engage with the country? And would the electric car revolution finally benefit the Congolese after years in which the wealth from its resources had rarely trickled down to the population? The Congo has provided the resources we have needed for over a hundred years, from the rubber for the advent of the motor car, to the copper for bullet casings in the First World War, and the uranium for the atomic bombs dropped on Hiroshima and Nagasaki.

A few months before the Wolfsburg meeting, employees of a German luxury carmaker had taken a trip to the Congo, a country of eighty million people, to show them where their cobalt came from. The Germans returned horrified by the conditions they saw at the small-scale mines in the south of the country, where young men and women dug for cobalt by hand for a few dollars a day with no safety equipment. There were tunnels full of young miners working below seventy metres when they were only supposed to be at thirty metres. Miners with no shoes were handling huge bags of cobalt. This was different to the well-ordered steel and aluminium plants they were used to buying metal from. 'They arrived with seventy questions and left with a hundred,' one guide recalled. The sheer number of people mining by hand and the poor working conditions were not what the polite Germans expected when they launched a drive to end the era of fossil fuels with the slogan 'the future has begun'. The future was arriving, yet it seemed to be coming out of Central Africa hitched to the practices of the past.

8

The Rise of a Cobalt Giant

'Merchants have no country. The mere spot they stand on does not constitute so strong an attachment as that from which they draw their gains.'

Thomas Jefferson[1]

'When you drive an electric vehicle we are a part of the journey.'

Glencore Nikkelverk[2]

A few months after the meeting in Wolfsburg, Ivan Glasenberg arrived at the Beau-Rivage hotel next to Lake Geneva in Lausanne, Switzerland, to deliver a stern warning. The lake glistened in the sunshine as the chief executives and traders, nearly all men in suits, gathered inside. It was a familiar environment for Glasenberg, an energetic billionaire South African who thrived on competition and doing deals. Security prevented protesters from Swiss NGOs from getting anywhere near the delegates, and they were forced to hold their protest on the road by the lake. It was here that the ailing kleptocrat Mobutu Sese Seko had fled during the final days of his thirty-year rule over the Congo, then known as Zaire, as the country collapsed and burned around him. Glasenberg's company Glencore was now set to make a fortune from Mobutu's country, thanks to the two large copper and cobalt mines it owned. After the failed talks with Volkswagen Glencore had upped the ante: striking a deal to sell most of its future cobalt production to a

Chinese battery materials company, GEM Co, which was based in the southern Chinese city of Shenzhen. Glencore had also previously sold cobalt to China's largest battery maker, CATL. If Western carmakers wanted cobalt from the Congo, they had better act quickly, Glencore was indicating. It was a canny move that was typical of Glasenberg, a born deal maker who had built Glencore into one of the world's largest mining companies through aggressive acquisitions. 'The motor car industry hasn't woken up to the fact, I don't think, how important cobalt is,' Glasenberg said in his slightly hoarse voice at the *FT* Commodities Global Summit, in one of his very rare public appearances. 'The Chinese will have most of the off-take of cobalt. They're not going to sell batteries to the world, more than likely they'll produce batteries in China and sell electric vehicles to the world.' Glencore would have no qualms about selling its copper and cobalt assets to a Chinese buyer if they offered a high enough price, Glasenberg added. 'They may come with a number that blows the lights out. If we get that number I've got to look after shareholders. I'm not here to look after the world politics, so we would sell it, yes,' he concluded.

It was well pitched – aimed at the heart of the insecurities many Western carmakers had about China, which were built around growing fears that they would lose out in the world's largest car market to their Chinese rivals after decades of efforts to enter the country. Part of the advantage for Chinese makers of electric cars was the easy access they had to a domestic supply chain of batteries and battery materials – as well as generous subsidies. Glencore was not the kind of company to take sides, Glasenberg was indicating; it was focused on making money. Cobalt prices had already more than doubled, earning Glencore millions more dollars on every tonne they sold from the Congo.

But Glasenberg's confidence actually masked a deep unease. Even as he was speaking in Lausanne the company's business ties in the Congo were coming under increased scrutiny, raising deep questions about how Glencore – a Switzerland-based mining company listed in London – had acquired its mines in one of the world's poorest countries. It was a past that threatened the company's position as a key supplier to the electric car revolution.

*

Based in the small Swiss lakeside town of Baar, a thirty-minute train ride from Zurich, Glencore is not the kind of company that clean energy advocates like to think of when they describe the future. It is one of the largest producers of coal from mines in South Africa and Australia, a fuel that is a key contributor to climate change. In the summer of 2018 at a mine outside Johannesburg I watched as black lumps of coal were dug up by a building-sized dragline excavator operated twenty-four hours a day, itself consuming huge amounts of coal-fired power from the grid. Trucks constantly drove around the mine site dispensing water to try to tamp down the abundant coal dust. The coal was exported from South Africa's Richards Bay to India, Pakistan and other countries in Asia which would burn it in power stations. But at the same time Glencore is also a big producer of cobalt, nickel and copper – all metals that will be needed in clean energy technologies from batteries to wind turbines. It's also one of the largest recyclers of electronic waste. With 158,000 employees and mines in over fifty countries across the world, Glencore is the hidden hand behind the global economy whose profits topped $9 billion in the first half of 2021 alone.

The son of Lithuanian immigrants to South Africa, Glasenberg grew up in a cosy suburb of Johannesburg during the height of apartheid. Physically fit and competitive he became a champion speed-walker but missed out on his chance to enter the Olympics due to international opposition to the country's racist policies. After studying for an accountancy degree in South Africa and doing an MBA at the University of Southern California he joined Glencore's predecessor company Marc Rich & Co. as a coal trader in 1984, a year after its founder had been indicted by the US Department of Justice (DoJ) on charges of tax evasion, wire fraud and racketeering. Rich, who became known as the 'king of oil' for his ability to strike deals selling oil from the Middle East outside the control of the giant Anglo-American oil companies, paid little heed to politics or the whims of the liberal elite. His company traded with whoever had raw materials to sell, including Iran under the Ayatollah and then South Africa under apartheid. The racist white regime in Pretoria may have been 'the single greatest source of his wealth'.[3] Rich saw corruption as a cost of doing business in many countries. 'The bribes were paid in order to do the business at the same price as other people were willing to do the business,' Rich once said.[4] A US Congressional investigation described Rich's business as 'based largely on systematic bribes and kickbacks to corrupt local officials'.[5] To Rich the only risk was if someone was unable to pay. It was a culture that worked well in the era of ever greater globalisation in the 1970s and 1980s. Large oil-producing countries no longer wanted to be controlled by the big Western oil companies, and wanted to sell their oil for a higher market price. Rich became the middleman par excellence, helping to cut deals across the globe.

Glasenberg thrived at Rich's company after joining as a young trader in the 1980s – personifying the hard-working, willing-to-travel culture that Rich had cultivated. Working first in South Africa and then Australia, Glasenberg also had a stint in Hong Kong and Beijing in 1989, during the Tiananmen Square protests, where he stayed in the Holiday Inn Lido and was evacuated from the city after the 4 June military crackdown. He became head of the coal division the following year. His big break came during a management-led buyout that ejected Rich from the company in 1993 following a failed bet on zinc prices. It was rechristened Glencore. Glasenberg ascended swiftly to the top spot in 2002, just in time for the world's largest commodity boom led by China. It was a boom unlike any Glasenberg had experienced before. It also led directly to the heart of the Congo.

*

In the Congo, the deep open pit of the Mutanda mine, the world's largest cobalt mine, lay below me in the morning sun, quiet and still. Its terraced slopes were neatly cut, ordered like Chinese rice paddies. A lone truck moved along the bottom of the pit, looking as small as a car, though it was in fact eighteen feet tall with massive diesel engines. Behind me the mine site looked like a small chemical industrial facility in China, with arrays of conveyor belts, pipes and tanks where copper rock was crushed, leached and processed to produce large dull-orange rectangles of metal that hung like clothes in a dry cleaner under a large hangar. The stiff copper was bundled together into packs that were placed on waiting lorries, which then drove to the port of Durban in South Africa. The cobalt was processed separately and dried into a dark green powder, cobalt hydroxide, that was bagged up and placed on different trucks driving the same route. The entire electric car industry was reliant on dozens of trucks that left the middle of Africa every day, driving along perilous roads, across multiple borders, to the coast.

Sulphuric acid trucks came day and night in the other direction to feed the mine's insatiable demand for acid. A few months before my visit in February 2019, eighteen people died when a truck carrying acid to Mutanda overturned. But the mine could not stop for any individual fatality. It ran twenty-four hours a day all year round, guarded by heavy security. Dogs milled around the entrance to the mine site, before they prowled the perimeter for intruders, especially local miners who tried to break in and mine the rich ground themselves.

Eric Best, an American executive who came from a family of copper miners in Arizona, placed a small bottle of the finished cobalt product on the desk in the quiet Mutanda office, which was set back behind a nicely manicured lawn. I stared at the fine dark green powder: the key for every electric car battery, the new prize for miners. It looked more like a colourful powder in a school chemistry set. Best opened up PowerPoint, but showed little enthusiasm for delivering a presentation. The mine had been besieged with so many visitors over the past few years due to the cobalt boom: NGOs, car companies and Glencore management. Mutanda was a geological rarity in the area: a mine spectacularly rich in cobalt. Its ratio of cobalt to copper was two to one, whereas most mines in the Congolese Copperbelt were ten to one, Mark Kenwright, one of the original geologists who helped examine the deposit for Glencore, told me. After rushing through his presentation Best asked if I had any questions. But there were rules, of course: one thing he didn't want to talk about was the mine's history, or how Glencore got this rich deposit.

Yet the history is evident at the mine site. On the rim of the mine a few lone trees stood silhouetted starkly against the brilliant blue sky. Over to one corner I saw some boulders and one of the distinctive orange tarpaulins found in the area that signify some-one mining for cobalt by hand. 'That's where the original artisanal mine was,' a young bearded Englishman who works at Mutanda told me, using the term for illegal hand diggers. In the beginning there was no vast hole in the ground or industrial mining site, just groups of local diggers mining and washing cobalt and copper as they had done for thousands of years. It was a telling remark, reflecting the deeper story of the Congo, where rich deposits of copper and cobalt have been transferred to foreign mining companies, who pay millions of dollars in taxes and royalties that get sent to Kinshasa, the capital, over eight hundred miles away. The money often ends up in the pockets of the political elite and little of it ever comes back, even though natural resources are the country's main export.

It was easy to forget all this as we visited the canteen for lunch. We drove along large mud roads, past the trucks that dispensed water to keep the dust down. It was a beautifully clear sunny day and I could see the green low-lying hills in the distance. The canteen was set back behind a lovely garden with neatly cropped trees. It was warm outside and the buffet was plentiful – there were large salads, multiple cheeses and colourful desserts. Miners with their families and girlfriends arrived – it was a multinational crowd, with Filipinos, Americans and South Africans. I watched a young white couple roll up in their Land Cruiser and hop out into the sun, looking like an advert for Timberland. It was a world away from the dust and crowded roads outside the mine site. I thought of the carmakers coming to the mine site and how impressed they must be with the order, efficiency and cleanliness. Tick, tick, tick, I imagined them thinking of their supply chain audits.

But as we left the mine that day, it was clear something was wrong. Outside the gates of the mine, we saw a young man, who was carrying cobalt ore in sacks on the back of his bike, being beaten and thrown to the ground. Our driver, who lived and worked for Glencore in Kolwezi, visibly shuddered at the scene, and said she would report it to the company. A few days earlier the Congolese army had shown up at the nearby Tenke Fungurume copper and cobalt mine, which is owned by a Chinese mining company, and she wondered if a crackdown was ongoing. The next day at breakfast at our hotel, she was suddenly called away for a meeting with the local governor.

The same day I was sent photos on WhatsApp from a local contact who said that hundreds of individual miners had invaded Glencore's other mine in Kolwezi, the Komoto copper mine, to mine for copper and cobalt. The photos showed hundreds of young men in dark clothing scrambling up and down a rampart of loose earth at the edge of a mining pit. The miners were so desperate for money that they had invaded an industrial mining site to dig on the earth – which Glencore considered its own property. A few days later the news came in that a total of thirty-six miners had died at the mining site after a landslide and I was sent more photos of dead bodies laid out on the ground, and miners trapped beneath rocks. It was a stark indication of how difficult it was for the company to isolate itself from the country where it generated its riches.

*

As I visited Mutanda, millions of dollars in royalties from the mine were not going to the ordinary Congolese but under contractual arrangements to one man: an Israeli billionaire called Dan Gertler, who had made more money for himself and Congolese elites than anyone since perhaps Belgium's King Leopold II. Gertler, a long-time friend of President Joseph Kabila, was the one of the keys to Glencore's success in the Congo, helping the company secure and operate its mines. Glencore lent him millions of dollars and in return Gertler helped with the company's government relations. For Glencore, Gertler was a middleman between government officials and the company; he was an 'arm's length' agent. Over ten years, Glencore gave more than half a billion dollars to offshore companies owned by Gertler, in the form of loans, shares and cash, allowing him to make at least $67 million in profits, according to the NGO Global Witness. A study overseen by Kofi Annan in 2013 estimated that five mining deals involving Gertler (including some with Glencore) saw the Congo lose out on $1.4 billion – twice the country's annual spending on health and education. The US Treasury would later put the figure at $1.6 billion; an anticorruption organisation later estimated an even higher figure of $1.95 billion.[6] Gertler has not been convicted of any crime and has denied any wrongdoing in his dealings in the Congo. He has said that he has 'always sought to operate in respect of the law and the interests of the people of the DRC'.

Gertler had grown up in a well-heeled neighbourhood in Tel Aviv, in a family who had made their money in the diamond business – polishing and cutting rough diamonds to be made into jewellery. He was the grandson of Moshe Schnitzer, one of the leading lights of Israel's diamond community, who had established and led the Israel Diamond Exchange from 1967 until 1993 and served as the president of the World Federation of Diamond Bourses. The family's other passion was football: Schnitzer was a keen football fan and Gertler's father was a goalkeeper for the Maccabi Tel Aviv team in the 1960s. Gertler had a typical Tel Aviv childhood with his three sisters, and played football enthusiastically for the local team. A talented footballer, his potential career ended after he suffered two head injuries while playing.[7]

Gertler wanted to branch out on his own rather than follow in his grandfather's footsteps in Israel. After serving in the army as a computer operator, he studied business administration and worked in the family's polished diamond business. He learned all about how diamond polishing worked. But he quickly tired of Israel's cosy diamond community and was drawn instead to the raw, uncut yellow diamonds that had arrived in Israel from around the world, which seemed to hold more mystery and adventure for him. 'He showed a great interest in rough diamonds,' Asher Gertler, his father, recalled. 'He learned all the stages of the cutting work. He just fell in love with rough diamonds.'[8] He also broke with his family by turning ultra-Orthodox.

He decided to source his own rough diamonds from Africa, founding his company Dan Gertler International when he was just twenty-two. He had an appetite for adventure. He first went to Liberia and Angola, buying up diamonds and selling them in the major diamond-consuming countries, coming into competition with some of the biggest players in Africa's diamond industry, which had been dominated for most of the twentieth century by De Beers. He was soon described as a rising star in the diamond industry: 'Bold, sophisticated, brutal...an adventurer with a short fuse'.[9]

Gertler then turned his attention to the diamond-rich Congo. He had come to the Congo earlier than Glencore. Many miners had rushed to sign contracts with the rebel leader Laurent Kabila in 1997 before he had even got to Kinshasa and deposed the aged dictator Mobutu. That same year a twenty-three-year-old Gertler stepped off an aeroplane in the Congo's dusty capital to seek his fortune. Kabila, a rebel leader who had been living in Tanzania since the 1970s, had just taken power and Mobutu had fled to Morocco, where he would die of prostate cancer later that year, after thirty-two years in power. Regime change is the ultimate opportunity for making money in natural resources.

Gertler had picked a moment when the Congo was at its most fragile and worn out. The country's entire economy needed rebuilding after years of neglect and theft under Mobutu, who had spent millions on a vast palace and airport in the remote town of Gbadolite and even encouraged officials to 'steal cleverly, little by little'.[10]

In the summer of 2000 Gertler visited Kabila at his office in the presidential marble palace in the capital, looking to strike a diamond deal, according to a lawsuit filed by Yossi Kamisa, a former Israeli policeman who later fell out with Gertler, in 2004.

The young Israeli promised to raise $20 million for Kabila in return for a near monopoly on sales of the country's diamonds from the state-owned producer MIBA. Kabila needed to raise money to pay back Rwanda which had helped him seize power, and oil and diamonds were some of the quickest ways to raise funds. When Laurent Kabila had first arrived in Kinshasa, he was broke, so one of his government's first stops was the Central Bank, where they found the huge cement chambers empty, with just a fifty French franc note left.[11] Shortly after coming to power, he was forced to fight a new uprising in the east of the country that had started in 1998, in what is known as the second Congo war, after his former backers, Rwanda and Uganda, turned on him.

Then in January 2001 one of Kabila's teenage bodyguards entered his office in the presidential palace and fired his pistol at the president. He died within hours.[12] The Congo was once again on the brink of chaos.

Gertler lost his diamond monopoly but continued to fly regularly between Tel Aviv and Kinshasa and began to ingratiate himself further with Kabila's son, the thirty-year-old Joseph, who took power after the death of his father. 'I have known Dan Gertler since 1997,' Kabila told the journalist Tom Wilson in an interview in 2018. 'He came, wanted to do business in the Congo, has been doing business in the Congo [and] definitely we want him to continue to do business in the Congo.'[13]

Kabila's position within the Congo was incredibly weak. He spoke bad French and was very young. So he sought legitimacy outside the country instead, hoping it would reflect well on him at home. In this Gertler was key. He became an 'ambassador-at-large' for the Congo, eventually taking Congolese nationality and driving around Tel Aviv with Congolese diplomatic licence plates. Along with his sidekick, Brooklyn-born Chaim Leibovitz, he also won favour with Kabila for helping broker a meeting in the Oval Office in 2003 with President George Bush.[14] Gertler also flew with Kabila to China,[15] where he had studied at the People's Liberation Army's National Defence University after his father had come to power. He became firm friends with the president, attending his wedding in 2006 in the garden of Kabila's riverside palace.

Following the signing of a peace deal to end the second civil war in 2003, the Congo's economy was badly in need of funds. The country's copper and cobalt mining industry had fallen on hard times, due to mismanagement by Mobutu and a collapse in copper prices in the 1970s. Equipment was rusted and abandoned, and laid-off mine workers scavenged by hand on the unguarded deposits. Utilisation at the mines was less than ten percent.[16] In 1990 one of the crown jewels, the underground Kamoto mine (part of the Katanga Mining complex which Glencore would later come to own) collapsed, leading to a twenty-three percent drop in overall production.[17] It later flooded. Gécamines, the state-owned mining company that was the former Belgian monopoly Union Minière, was a shell of its former self.

But in 2005 miners around the globe were scrambling to secure copper, as prices rose to fresh records because of China's insatiable demand. At the same time Kabila was preparing for the Congo's first elections in 2006, and needed money. Kabila allowed Augustin Katumba Mwanke, a former banker, to raise funds by agreeing to new joint ventures. 'Short, unpretentious and polite,'[18] Katumba had left the Congo (then Zaire) as a young man to study and work in South Africa, where he was a banker for HSBC Equator Bank. He met Laurent Kabila during the early days of the rebellion while on a business trip, and was offered a position in the Ministry of Finance. In 1998 Kabila then made him governor of Katanga province, where the large copper and cobalt mines were. He became close both to Kabila's family as well as various mining interests. When Kabila's son Joseph took power Katumba was given the job of managing state assets; he was also in charge of international mining investments. He became one of Kabila's closest advisers, and the *FT* dubbed him the 'Dick Cheney of the Congo' – the power behind the throne, who had a 'casual grandfatherly stature' and looked like 'the chauffeur in *Driving Miss Daisy*'.[19]

Katumba became close with Gertler in what became known in later years as a 'mousetrap'. 'Every company that wanted a major mining deal since 2008 had to enter the mousetrap and "get caught"/pay Gertler et al.,' one former diplomat told me. Katumba was a master of the *envelopperie*, or practice of receiving or giving bribes using cash in an envelope, according to Congo expert Jason Stearns.

As Kabila approached his first election in 2006 Katumba negotiated a series of mining deals that ended up transferring ownership of some of the best mines to private investors, including Gertler. These were then sold on to major mining companies for considerably higher prices. Even at this early stage the deals were criticised by the World Bank, which said thorough valuations had not been conducted by Gécamines and the negotiations had taken place with a 'complete lack of transparency'.[20] 'Warning! A fire-sale!!!' reported a Congolese anti-corruption organisation.[21]

It was around this time that Gertler became Glencore's key business partner in the Congo, a relationship managed personally by Glasenberg. In a few short years Glencore ended up owning two of the Congo's best copper and cobalt mines, beating out rivals and other foreign adventurers.

The Mutanda mine had begun as a clearing nestled on the side of a hill where locals came to dig for copper and cobalt. Known as 'barefoot geologists' for their skill in finding rich deposits, a concentration of local miners on one spot is a sure sign the ore is rich with minerals. By the early 2000s the ore was being traded by a local Lebanese businessman and Kolwezi resident Alex Hamze, whose Groupe Bazano ran a trucking and logistics business in the Congo. Cobalt and copper rocks were washed, bagged and trucked out from the site – some of it reaching Glencore's smelter in neighbouring Zambia. Impressed by the high grade of cobalt, Glencore sent geologists to drill the site, where they stayed in conical huts and ate Lebanese food provided by Bazano. Very quickly it was clear that Hamze was sitting on top of a mother lode: Glencore took a forty percent stake in the project and started to build an industrial mine.

A year earlier in 2006 Gertler had obtained from Gécamines a stake in a neighbouring deposit to Mutanda, a 185-square-kilometre concession called Kansuki. In 2011 just as the Mutanda mine started to properly produce copper and cobalt, Glencore revealed that an offshore company linked to Gertler had also bought the twenty percent stake in the Mutanda mine owned by Gécamines, the Congo's state-owned mining company. Gécamines had sold it (as well as its twenty-five percent stake in Kansuki) to Gertler for $220 million, including a debt of $31.4 million. Yet Glencore's IPO prospectus suggested the Mutanda mine alone was worth $3 billion, suggesting Gertler had got an extremely good deal indeed.* In 2013 Glencore then merged Mutanda with Kansuki. The deal enhanced 'the successful relationship that has existed for many years between the two companies', Gertler said.[22] But Glencore had spent $1 billion building Mutanda, while Gertler said he had spent just $100 million exploring Kansuki. With the two merged Gertler was able to receive millions of dollars in royalties and payments from what would become the largest cobalt mine on the planet.

Gertler and Glencore had won over the early rough-and-tumble adventurers who had come to the Congo. They were no match for Glencore's firepower. Hamze of Groupe Bazano ended up being bought out and retiring to Malta to enjoy his riches. Glencore's power helped Gertler survive, according to the NGO Global Witness. 'With the London-listed giant behind him, he was able to knock out the most powerful men in Congo's mining sector,' the group said.[23]

---

* Gertler has said that the valuation was 'subjective'. See 'Responses by Dan Gertler to Global Witness', https://cdn2.globalwitness.org/archive/files/library/responses%20by%20dan%20gertler%20to%20global%20witness.pdf.

Once it had secured its two mines Glencore was well positioned in the Congo. Copper prices had recovered from a financial crisis-induced downturn and were heading towards a record $10,000 a tonne. China was consuming around half of the world's commodities – an unprecedented amount for any country. All mining companies had to do was dig metals up and send them on a boat to China. China wanted everything: copper, aluminium, iron ore and cobalt. Glencore, which both mined and also traded commodities, was in pole position to benefit, especially since its £10 billion London listing in 2011 which had given the company firepower to buy and build copper mines. Glasenberg was on the hunt for deals and Glencore were the smartest guys in the industry. The company bought up coal mines in South Africa and in early 2012 took over its long-term rival Xstrata, adding a considerable number of mines to Glencore's balance sheet. Glencore had now become a giant to be reckoned with.

The stock market flotation had also made billionaires of Glencore's key executives, including Glasenberg, as well as the company's head of copper, Aristotelis Mistakidis, who oversaw the Congolese assets. Razor-sharp and quick-witted, Telis, as he was known, had an in-depth knowledge of the global copper market and could regale dinner guests with details of every copper mine in the world. He became known as the 'copper king', the man who ruled the global copper market. Mistakidis grew up in Rome, where his father worked for the United Nations as a marine biologist, and later studied at the London School of Economics. He took no pains to hide his newfound wealth. In 2015 he bought a duplex in a block designed by Norman Foster in Belgravia, one of London's most exclusive districts, for £46 million.[24] At 7,900 square feet, the sale was the third-highest price for an individual flat in London at the time. It also marked the peak of the global copper market. The following year, copper prices crashed to their lowest levels since the global financial crisis.

Glencore's problems started from an unlikely source: one of New York's most successful hedge funds: Och-Ziff.

In February 2012, a young hedge fund manager in London, Vanja Baros, sent an email to the head of the London office, an American called Michael Cohen, copying a *Financial Times* article on the death in a plane crash of Kabila's aide, Katumba. 'FYI', it said, pointing out he was the 'key guy' of their DRC partner, who matches the description of Gertler. A few days later Gertler sent a text to Baros which said: 'I'm fine ... sad but fine ... I will have to help Kabila much more now ... tomorrow the burial will take place.'[25]

Two years later, in 2014, Och-Ziff announced it was under investigation by the US DoJ and the Securities and Exchange Commission on charges of bribery.

Och-Ziff had been founded in 1994 by Daniel Och, a former Goldman Sachs banker, to manage the Ziff publishing fortune. The hedge fund, which had $39 billion under management, was run in London by Cohen, a partner who had been sent to set up the London office in 1999 at the age of twenty-seven, having joined two years earlier from Franklin Mutual Advisers. Talked of as Och's protégé in one court testimony, Cohen was the third best compensated executive when the company went public in 2007.[26] The same year Cohen bought a Hampshire estate owned by the Duke of Wellington for £12 million.

But China's commodities boom and the potential money to be made sucked Cohen into Africa where, in late 2007, just as Glencore was entering the Congo, Och-Ziff held discussions with Gertler about entering into a partnership 'based on special access to lucrative investment opportunities in the DRC involving the country's diamond and mining sectors'. An Och-Ziff background check on Gertler found that he 'is happy to use his political influence against those with whom he is in dispute ... [and] keeps what can only be described as unsavory business associates', according to the DoJ's settlement which does not mention Gertler by name but calls him an 'Israeli businessman'.[27]

Starting in 2008 Och-Ziff entered into several DRC transactions with Gertler, with the 'understanding that Och-Ziff's funds would be used, in part, to pay substantial sums of money to high-ranking DRC officials to secure access to, and preference for, the investment opportunities'. In late 2008, an Och-Ziff employee was alerted that an audit of Gertler's records revealed payments to DRC officials, but the employee instructed that any references to those payments be expunged in the final audit. The DoJ said that between 2005 and 2015 'An Israeli businessman…paid more than one hundred million US dollars in bribes to obtain special access to and preferential prices for opportunities in' Congo's mining sector. William Sweeney, assistant director in charge of the FBI's New York field office, called it 'bribery in its purest form'.[28] One London-based NGO, RAID, calculated that Och's personal fortune would last the lifetimes of more than 95,000 Congolese based on the $1.25 each day that most Congolese live on. Gertler has not been convicted of any crime and has denied paying bribes in the Congo. Och-Ziff settled the case against it in 2016 and agreed to pay $412 million in criminal and regulatory penalties.

Publicly, Glasenberg remained defiant about Glencore's connections with Gertler. In August 2016 he dismissed concerns about Gertler and the DoJ probe into Och-Ziff. 'We've heard nothing,' he said. 'Dan, you'll remember, he's a partner in Mutanda. That occurred because he had an asset next door which we merged together with Mutanda.'[29] But the DoJ settlement must have made nerve-wracking reading for Glasenberg. It alleged that money had been paid in bribes to Congolese officials.

Glencore liked to portray itself as a company of action. It moved quickly to buy out Gertler from its mines in the Congo, believing that it could distance itself from him. In February 2017 it reached a deal to pay him $534 million, after all Gertler's debts to Glencore had been repaid – for his 31% stake in Mutanda and 10.3% in Glencore's other mine, Katanga. Investors breathed a sigh of relief, and shares in Glencore rose to their highest levels since November 2014. That November Mistakidis attended the annual event of the LME, sipping espressos and meeting his usual clients at the 5 Hertford Street private members' club in Mayfair.

Gertler would still receive royalties from the mines, but he was gone as a partner. Glencore prepared to position itself to supply the world's electric carmakers with cobalt. In 2017 Glencore started negotiations with Tesla, Apple and Volkswagen, taking Tesla to the DRC to visit its mines. Glasenberg said publicly that he didn't believe that the world could produce enough cobalt to satisfy long-term electric car demand.

But pressure had been growing in Washington and Europe to isolate President Kabila and force him from office. Kabila had overstayed his welcome: there were protests in the streets, and he had tried to thwart the constitution by lingering beyond the official end of his second term at the end of 2016. (The Congolese Constitution does not allow presidents to run for a third term.) He had sold some of the country's key mines not only to Gertler but also to the Chinese. In June 2017 the US Treasury imposed sanctions on one of Kabila's most senior military officials, General François Olenga, as well as US visa bans on some of Kabila's family. The US ambassador to the United Nations Nikki Haley visited Kinshasa in October 2017 and read the riot act to Kabila at a ninety-minute-long meeting at his waterside palace – saying he must hold elections in 2018 or the US would withdraw all its support to the country. Then in December, the US Treasury took aim at Gertler, under the Magnitsky Act, which had been passed by Congress in 2012 to sanction Russian officials who had been involved in the death in detention of the tax advisor Sergei Magnitsky. The sanctions immediately locked Gertler out of the

US banking system, also making him unable to visit the US, or engage in any transactions with a US company. Any company outside the US also faced the risk of being sanctioned by doing business with him – putting Glencore in the direct firing line. Glencore, which sold all of its commodities in US dollars around the world, had no choice but to halt payments to Gertler, which they did immediately after the sanctions.

In April 2018 local staff at the gates of the Mutanda mine received an unsettling visit from local Congolese officials who said they were acting on behalf of a local court in Kolwezi. A bailiff entered the mine and put Post-it notes on all of the office equipment, to designate what the local court was going to seize; they went outside to the copper processing circuit and did the same. They were acting on behalf of a lawsuit filed by Gertler, who was seeking $3 billion in damages for unpaid royalties. Gertler had filed a lawsuit in the Kolwezi court that authorised the bailiff to freeze bank accounts, assets and titles over mining assets up to $695 million from Mutanda and $2.28 billion from Katanga. Gécamines had filed its own lawsuit in the same court against Glencore that sought to dissolve Katanga Mining due to its high level of debt. Glencore's adventure in the Congo was beginning to look more like a nightmare than a daring and brilliant move by Glasenberg.

Glasenberg and his head of strategy, Oxford-educated Paul Smith, a former banker, came up with a plan. They flew to Washington and met with officials in the Trump administration, including the Department of Defence, the Treasury and the State Department, telling them that if Gertler sold their two mines, the only buyer would be a Chinese company – giving China a one hundred percent share of the global cobalt market just as electric cars were taking off. Did the US really want to be held hostage to China for every EV it wanted to make? The US officials gave tacit support to Glencore's proposal, according to company officials. Glencore felt confident that its message had been received and its point taken – especially in a Trump administration that was growing more concerned about China's control of mineral resources and its influence in Africa. Glencore officials also briefed the Swiss authorities on the same plan, the company said. 'Gertler effectively said to them I'm going to take you to court and I'll probably beat you in court and when I beat you I can effectively get the mine and sell to someone else – that someone else is going to be Mr China,' a person familiar with the event said.

Glencore decided it would pay Gertler his royalties in euros. Their legal advice showed that they could get around the sanctions by paying a non-US individual in a currency other than the dollar. Glencore had 'carefully considered its legal and commercial options', they said, and this outcome would 'appropriately address all applicable sanctions obligations'. As part of the deal, Gertler agreed to withdraw all litigation in the courts. 'Katanga believes that payment in non-US dollars of royalties to Ventora [Gertler's company] without the involvement of US persons would appropriately address all applicable sanctions obligations,' the company said.[30]

Glencore's more immediate problems in the Congo did recede. The same month Glencore resolved its dispute with Gécamines over Katanga Mining – agreeing to wipe off $5.6 billion in debt built up under Mistakidis and pay Gécamines $150 million relating to historical commercial disputes as well as $41 million to cover expenses incurred as part of an exploration programme. Glencore had manoeuvred its way out of trouble.

But some in Washington did not see things the way Glencore did. The same day Glencore decided to pay Gertler in euros, the DoJ sanctioned fourteen of Gertler's offshore companies – including Ventora, the vehicle for Glencore's royalties from the Katanga mine.

Less than three weeks later Glencore revealed that the DoJ was investigating the company for its activities in the Congo, Venezuela and Nigeria. Glencore's shares fell by as much as twelve percent, cutting $5 billion from its market value. The DoJ's subpoena requested documents going back to 2007 – the fateful year when Glasenberg decided to enter the Congo to mine the gentle hills of Mutanda. Glencore appointed Tony Hayward, the former head of BP who had alienated US authorities during the company's oil spill off the coast of Louisiana, as head of the committee responding to the DoJ. Trim and athletic Hayward embraced the Glencore boys' club, and could be quick to anger when asked by investors why Glencore had not distanced itself from Gertler. Hayward had told the NGO Global Witness in 2014 that 'there is absolutely no basis for any investigation and we have no intention of carrying one out'.[31] Now four years later he was in charge of providing the paperwork for just such an investigation. In May 2022 Glencore pleaded guilty and agreed to pay $1.5 billion to settle investigations by authorities in the US, UK and Brazil. Glencore admitted that it paid $27.5 million to agents in the DRC, 'while intending for a portion of the payments to be used as bribes to DRC officials, in order to secure improper business advantages.'

At the same time Congolese officials could see the EV revolution coming and wanted a piece of it. The Congo was well aware of the rise of electric cars and the role it could play in supplying them. It produced over seventy percent of the world's cobalt – giving it a monopoly much in excess of that enjoyed by Saudi Arabia in oil. In January 2016 I had met the chairman of Gécamines in the hushed and opulent surroundings of Claridge's hotel at the heart of super-rich London. King Leopold II had stayed there in 1876 when he visited the city to rally support for an expedition to the Congo. Dressed in a tidy three-piece suit Albert Yuma was flanked by aides and sat comfortably back on a leather sofa. His hair was combed impeccably – he could have been a wealth adviser talking to the uber-rich about their tax strategies. Instead he held the key to the resources of one of the poorest countries in the world, which had all the cobalt needed for a shift away from fossil fuels. 'You don't make electric batteries with dollars but with cobalt from the DRC,' Yuma liked to say. 'As the Congo – we want to take the opportunity – like the Arab countries took the opportunity of oil to build their country and to share the revenue with their population,' Yuma told me. When I attended the DRC Mining Conference in the dusty town of Lubumbashi, a couple of hours' drive from Kolwezi, in the summer of 2019, Yuma was greeted like a celebrity by local attendees. He repeated his insistence that Gécamines had failed to receive any dividends since the country introduced a mining code in 2002. 'We were told you will have milk and honey but twenty years later we have nothing,' he said.

Yuma's arguments resonated because international mining companies had for the most part been treated very leniently in the Congo. But in the eyes of foreign mining executives all Gécamines wanted was money – and lots of it. 'They are like Russian oligarchs, they don't know how much money they need, they just want more of it,' one foreign mining executive active in the DRC told me. Even if they paid more money in royalties and taxes, one foreign mining executive asked me, where would it all go? There would be no guarantee the money would go into spending on healthcare and education. Such sentiments were strengthened by stories of Yuma's lavish wedding ceremonies for his daughter in Kinshasa. The data itself also told a brutal story: annual income per capita in the Congo, at $785 in purchasing power parity terms, was still among the lowest in Africa. According to the Carter Center around $750 million that was paid to Gécamines from asset sales and royalties between 2011 and 2014 was missing from its accounts.*

---

* In November 2018 Yuma denied that vociferously – via an elegant-looking report that attacked all NGOs. Gécamines said all that money could be traced in its accounts, 'dollar by dollar,' he said. 'Their [NGOs'] only objective is the destabilisation of the DRC to serve without hindrance, in the name of pseudo-democratic ideals, foreign demand for cobalt, coltan, copper, gallium, germanium and other strategic minerals that the world so badly needs to ensure its energy transition,' Yuma said.

Despite the criticisms, however, Yuma was close to Kabila and momentum had been building in Kinshasa for a change in the country's 2002 mining code, which had been formed in the middle of a civil war – in order to secure more of the cobalt revenues for the Congo.

In early March of 2018 a number of small private planes landed in Kinshasa N'djili International Airport. They carried the chief executives of some of the largest miners in the country: Robert Friedland of Ivanhoe Mines, a legendary stock promoter who had been university friends with Steve Jobs from Apple; Glasenberg of Glencore; Mark Bristow, the tough-talking South African head of gold producer Randgold; and Steele Li, the head of China Molybdenum, which had just bought up one of the DRC's largest copper and cobalt mines. It was not easy to summon this group of leaders to one place, and only a few countries could manage it: China being one, and the DRC the other. They had been summoned last minute to see Kabila. Kabila was in no rush, however. The CEOs were left to stew in their hotel. The next day they were summoned. In an airy high-ceiling room Kabila sat on a chair like a throne in the middle of the room, and the miners to the side of him in a row. On the other side were Congolese officials, including Yuma. The miners urged Kabila to ease back on publicly released plans for the new mining code, which threatened to raise taxes substantially on international miners and declare cobalt a 'strategic mineral'.

Kabila listened patiently to their pleas, which were led by Bristow, who had long experience dealing with African governments. But Kabila seemed to know very few actual details about the new proposed mining code, according to people who were there. The meeting ended with no resolution.

Afterwards, Kabila signed into law a new mining code that gave the government the right to raise taxes to ten percent on minerals it deemed strategic 'on the basis of the Government's opinion of the prevailing economic environment'.[32] Cobalt was soon designated such a mineral. It also required that Congolese contractors should be used by mining companies and Congolese should hold at least ten percent of the shares of new mining ventures.

Some investors decided to exit Glencore, fed up with the company's ties to Gertler and its promotion of coal. 'Not only do Glencore perpetuate the slowing of decarbonisation of thermal coal they actively lobby against coal regulation in emerging markets,' one London fund manager said. 'Clearly, Glencore faces a tough dilemma. Either it could continue its business relationship with Mr. Gertler – now on the US sanctions list – to protect its valuable copper and cobalt assets, or it could end this relationship,' London fund Sarasin & Partners, which sold its holdings in Glencore, said. 'Aside from obvious questions over how Glencore will untangle itself from this Gordian Knot, a key question for shareholders is why Glencore has – according to material published by Global Witness – chosen not to take opportunities to distance itself sufficiently from Gertler since IPO.'

The Brussels-based NGO Resource Matters estimated that Gertler received an average of more than $200,000 a day in royalties throughout 2018 on sales of cobalt and copper. That cobalt was now making its way to some of the largest car companies in the world, as well as into our smartphones.

Two years later Tesla's Elon Musk opened the company's new Gigafactory on the outskirts of Shanghai, which had been built in just ten months. Sitting in the audience were two young men from Glencore. Tesla's Shanghai Gigafactory was a watershed for the car company, as it opened up the largest electric car market and also turned Tesla into a global company. Tesla deepened its relationships with key Chinese suppliers such as battery maker CATL and lithium producer Ganfeng. To increase its control over its own raw materials, Musk also agreed to buy cobalt directly from Glencore.

The two companies signed a multi-year supply agreement. 'We want to become Tesla's procurement arm,' one Glencore executive said later. It was one of Glasenberg's last coups before he stepped down as chief executive in 2021 after twenty years at the helm. It also put Glencore at the heart of the electric vehicle revolution.

*

At 5.30 a.m. on a cold, dark crisp morning in February 2019, Glasenberg waited in the empty well-lit lobby of the Four Seasons hotel by Hyde Park for his morning run. Alert and energetic, Glasenberg looked as if getting up at 5.30 was entirely normal. Such early morning runs had become a ritual for Glasenberg and an assorted group of hangers-on, every time he was in town to announce annual results. It was a chance to ask questions – if you could keep up.

A day earlier Glasenberg had announced a dramatic change in Glencore's culture and altered its future direction – repudiating the key thing he had built his career on: coal. Under pressure from the Church of England and a group of other large investors known as Climate Action 100+ with more than $32 trillion of assets, Glasenberg had committed to capping Glencore's coal production at its current levels, at 150 million tonnes a year, and not buying any more coal mines. That would decline as the mines depleted, Glencore said. A man who had led Glencore's coal business in Asia and Australia, building it up through acquisitions, had vowed to not make any more in coal. It also went against his beliefs that coal would still be needed in the future, in countries such as Pakistan, India, Malaysia and Vietnam. 'What transition?' Glencore's chairman Tony Hayward had asked at a *FT* conference the previous year in response to a question about the energy transition. The Glencore men told things as they were, not how they hoped them to be. It was an attitude similar to the first lines of V.S. Naipaul's book, *A Bend in the River*, about the Congo: 'The world is what it is; men who are nothing, who allow themselves to become nothing, have no place in it.' Now for the first time since Marc Rich founded the company, it was publicly making a commitment to the planet's well-being. 'The environment changed around Glencore,' Edward Mason, who led the charge for the Church of England, said.

At sixty-two, slowing down was not an option for Glasenberg. As the group headed across the road into the darkness of the park, he quickly picked up speed. Slightly hunched over, he ran in the manner of an ex-speed-walker, robustly and fluidly. He began to chide those, including me, who couldn't keep up. We ran on by the light of the bright moon, feeling his energy still undimmed.

9
Blood Cobalt

'No one wants to touch ASM [artisanal mining] with a bargepole.'

Head of procurement at a carmaker

'The economic history of Congo is one of improbably lucky breaks. But also of improbably great misery. As a result, not a drop of the fabulous profits trickled down to the larger part of the population.'

David Van Reybrouck[1]

In 2014 a policeman in the small town of Kasulo in Kolwezi in the Congo decided to dig for a new latrine for his family. As he dug a pit in his yard, he discovered a rich source of cobalt in the black earth. He started digging under his living room, his bathroom, his bedroom and his kitchen. Neighbours joined in and soon the area 'looked as if it had been bombed', according to journalist Michael Kavanagh, who visited there in 2015.[2] Even a local pastor dug a large hole in the church floor to mine for cobalt. The houses started to crumble and the streets were full of holes, as families started to dig for cobalt and sell it to local markets. 'You would come out of your house and there'd be a big hole,' said one Chinese executive. There were regular deaths and injuries, and the local road had to be closed after tunnels were dug into its foundations. A major landslide killed dozens. Soon hundreds of people were dying as tunnels collapsed on people. An official put the number of dead at 250. One of the companies buying cobalt from the area was China's Huayou Cobalt, which was founded in the county-level city of Tongxiang in eastern China by an entrepreneur with no schooling, Chen Xuehua. Huayou quickly became one of the biggest buyers of cobalt from around Kolwezi, through its local entity, Congo Dongfang Mining International (CDM), which it had founded in 2006. Unlike Glencore, which acquired large mines with the help of Dan Gertler, Huayou relied mostly on cobalt from individual Congolese miners. Like many buyers in the years following the official end of the Congo's wars, Huayou didn't pay attention to where the cobalt came from, or who was mining it. The cobalt was high quality and cheap, and did not require the large amount of capital needed to build a mine. Locals knew how to find the best deposits, or they would mine on the edge of old mining sites once run by the state-owned mining company Gécamines. They would then sell the material to local traders, who would then sell to Huayou, who would crudely process the metal and then send it by truck to the port of Durban and then straight to China where it entered global supply chains. By then the cobalt had been mixed with other sources and 'cleaned', just like dirty money, giving us no sense of where it came from, or the blood and sweat and labour of those involved.

Like international corruption the cobalt trade had worked seamlessly for years, providing the cobalt for the batteries in our smartphones. Since there was no armed conflict in Kolwezi, cobalt was not called a 'conflict mineral', unlike tin, tantalum and tungsten, which required disclosure under the US Dodd-Frank Act of 2010. The thousands of men, women and children who dug for cobalt around Kolwezi and frequently died were given little help from the electronics giants in Asia and the West who used their products. They represented the forgotten bottom nodes of global supply chains. By the time Huayou was supplying major smart-phone makers such as Apple, children represented an estimated up to forty percent of individual cobalt miners.[3]

They were children like Kongolo Mashimango Reagen, who used to spend his days carrying twenty-five-kilogram sacks of cobalt from small mines in a southern corner of the Democratic Republic of the Congo, which is so rich in minerals that large deposits can be found just metres below the surface. His days in Kolwezi would start at 5 a.m. and accidents were common as tunnels dug by hand in the bright red earth collapsed. Miners drank beer and whisky and smoked to get through the day, he recalls. His uncle sold the cobalt to local traders known as *négociants*, and Kongolo received free food and accommodation as his payment. 'It was very tiring, very difficult,' he told me, standing in the bright sunlight on the edge of a makeshift football pitch by a school in Kolwezi in the DRC's south-east. 'I watched too many collapses. I have seen children dying in the mines.' The seventeen-year-old escaped the mines and now attended the school with the help of Good Shepherd, a Catholic charity based in Kolwezi. He had since learned to read. In 2014 Unicef esti-mated that some 40,000 children worked in the cobalt mines in the Congo, a year when mobile phone sales topped 1.9 billion. As late as 2019, the Organisation for Economic Co-operation and Development (OECD) found that children were present or working at about one in four artisanal mining sites. One study based on surveys in the former Katanga province estimated that about twenty-three percent of the children worked in cobalt mining.[4]

This was the logic of global supply chains. Their complexity meant the true environmental and social costs were kept hidden. It was an example of how our continued consumption of goods imposed what Peter Dauvergne called 'ecological shadows of consumption' onto distant communities and poorer people.

These shadows have continued to grow even as global concern about the environment has increased – with cobalt being a classic example. We throw away our old phones to buy new ones, or exchange our old cars for electric ones – so that total cobalt demand has only kept rising, despite any improvements in the amount of cobalt used per device. 'Environmentalism has failed to slow the ways that producing, using, and replacing consumer goods deflect ecological costs into distant places and future generations,' Dauvergne wrote in 2010.[5] Global cobalt production rose from an average of 38,000 tonnes a year between 1970 and 2009 to around 145,000 tonnes per year between 2010 and 2019.[6]

The environmental and social costs of that increase – which weren't reflected in any consumer price – were borne by the people of the Congo. While the number of deaths and injuries are not known, studies are starting to show the serious health burden of cobalt mining. In Kasulo, a study carried out in late 2014 and 2015 by scientists at Belgium's KU Leuven and the University of Lubumbashi found that children's urine and blood contained high concentrations of cobalt and other metals compared to those living in a nearby control area. In addition, there was evidence of 'exposure-related' oxidative DNA damage which was most pronounced in children. The researchers collected blood and urine samples from seventy-two Kasulo residents, including thirty-two children. A control group with a similar composition was selected in a neighbouring district. According to Professor Benoit Nemery, a doctor-toxicologist at the KU Leuven Department of Public Health and Primary Care, the results of the study were worrisome: 'Children living in the mining district had ten times as much cobalt in their urine as children living elsewhere. Their values were much higher than what we'd accept for European factory workers. This study may be limited in scope, but the results are crystal-clear. The differences cannot be attributed to coincidence.'

The study's conclusion was damning: 'the currently existing cobalt supply chain is not sustainable.'[7] A separate study published in the *Lancet* in 2020 also found that children born in the mining town of Lubumbashi were more likely to have birth defects than those born outside, and those risks increased if their father worked in a copper and cobalt mine. While little is known about the effects of prenatal exposure on birth outcomes, the study was conclusive: 'Paternal occupational mining exposure was the factor most strongly associated with birth defects,' it said.[8] The mothers of both the case and control group were found to have 'metal concentrations that are among the highest ever reported for pregnant women'.

*

Travelling at 279 kilometres per hour from Shanghai on the high-speed train, in early 2019 I visited Huayou's factory in the town of Quzhou, which is well known for its chemical industry. I travelled with Bryce Lee, who had been given the new position of head of corporate social responsibility at the company. Lee joined Huayou in 2005 after teaching English to the chief executive's nephew in the company's headquarters in Tongxiang. He told me about his experiences of the Congo on the quiet train as we raced through the rain-soaked paddy fields of central China, showing me pictures on WeChat. It had not changed much since he first went, he said. The people had nothing but were always happy, unlike China where people always strove for more. There were factories dotted all around the vast train station when we arrived, despite Quzhou's position on a tributary of the Yangtze that is the water source for Shanghai. Lee and I were driven to Huayou's gates which were bustling with construction activity. There were new buildings covered in scaffolding, which was a new plant built with LG Chem of Korea to make battery cathodes from cobalt from the Congo.

Lee showed me around the factory and guided me through the different processing stages. I saw the open-air warehouse where large white bags of raw cobalt arrived by truck from Shanghai or Ningbo and were stacked ten bags high.

All of the cobalt from the Congo was mixed together, no matter whether it came from the individual mine sites such as Kasulo or the larger industrial mine sites that Huayou owned in the country. Inside the factory the cobalt was then crushed, pumped as a slurry to be leached with acid in large tanks, filtered, its impurities removed, then heated in furnaces to form a pure cobalt solution that was then mixed with nickel and manganese to make battery materials. The final cobalt solution had to be perfectly pure to prevent the risk of fires in the battery, and was examined at the level of the atom by machines. In one room there were rows of large steel tanks where the air was heavy with the smell of metals and chemicals, a smell so strong I felt light-headed. 'No matter where we invest, we have to be responsible to local society,' a slogan on the wall said. The factory had a capacity of 30,000 tonnes of cobalt a year, a third of the entire global market. It was one of the key links between the Congo and the electric car.

We walked outside to a separate part of the factory where large battery packs from electric buses were disassembled to recover the 'black mass' powder that contains lithium, nickel and cobalt which can be leached out. In China every battery pack is given a barcode and must be recycled according to law. Lee said that under President Xi Jinping environmental controls were 'tough', and his environmental campaign to create an 'ecological civilisation' was very strict. He told me carmakers are very keen on using recycled cobalt in order to avoid the Congo, and in 2019 over fifty had already signed up to buy recycled materials from Huayou. The process is separate from the rest of the factory, so as to assure carmakers that there is no mixing with material from the Congo, which might be the product of child labour. Huayou saw it as a huge growth area and in 2020 targeted around 3,000 tonnes of cobalt from recycled material. Buses provided the first batteries they could get hold of due to the intensity of their usage. Still, it was a fraction of the cobalt that carmakers needed, which meant Huayou was in the Congo to stay.

Rail-thin with a closely shaved head, Huayou's founder Chen Xuehua often slept and ate at the factory, and could talk about the technical details of cobalt processing for hours. Chen was born in Tongxiang at the beginning of the Mao era in China. At the age of fifteen Chen was forced to leave school and had to find a way to make money. His family raised chickens, ducks and rabbits but Chen chose to sell beansprouts. He would ride his bike eight or nine kilometres every morning to the local market. At the age of nineteen he started working at a local factory, but in the morning, he would continue to sell beansprouts before work. After the factory went bankrupt in 1994, he founded his first chemical company, which made nickel chemicals. 'It was very basic, just cooking the material like in a kitchen,' Lee said. He then started to move into making cobalt for ceramics, founding Huayou Cobalt in 2002. As the company grew it needed overseas supplies of cobalt, so Chen came into contact with international traders, who as usual added in their margin and were difficult to deal with. Chen noticed that all of the cobalt came from the Congo and started to think it would be easier to go there to secure it than buying from international traders. So in 2003 he went, just as the country was recovering from its second civil war.

It was a period when China was urging its companies to 'go out' into the world to secure resources and business. Backed by credit from state-owned policy banks, Chinese companies accelerated their move into Africa, building roads, ports and infrastructure and investing in mines. In one of the continent's largest deals in 2007 the DRC's President Joseph Kabila signed a $9 billion resource-for-infrastructure deal with China's state-owned China Railway Engineering Corp and state hydropower company Sinohydro. The Sicomines joint venture would invest $3 billion in a copper and cobalt deposit, and invest $7 billion in infrastructure in the Congo. Future mineral sales from the mine would pay back the investment. The deal was backed by the China Export-Import Bank, one of China's two policy banks. It was a classic move by what became known as 'China Inc.' that typified Beijing's strategy in the 2000s. Its size was larger than the Congo's state budget the year it was signed. The deal cemented China's relationship with Kabila, who, as we saw, had studied at the People's Liberation Army's National Defence University in February 1998, and gave the state access to Congo's resources.

At the same time under the surface of these large government-to-government announcements small nimble private companies like Huayou were spreading out across Africa and bringing Chinese staff to work in smelters and as traders. The Congo was an attractive location for small Chinese businesses as opportunities to buy cobalt were plentiful. Artisanal mining had been encouraged by Kabila's father Laurent Kabila after he came to power in 1997, and China had few domestic sources of the metal.* The Congolese mining code of 2002, which was drafted with the help of the World Bank, designated artisanal mining as legal in authorised areas, although few of these areas

---

* China's reliance on imported cobalt from the Congo is its greatest reliance on one country for any commodity, more than iron ore, which is mostly mined in Australia and Brazil.

were created or enforced. A year later the Congo's state miner Gécamines laid off 11,000 miners, many of whom went on to mine by hand, becoming *creuseurs*, since they had spent most of their pay already on cars and TVs.[9] Some of the old mines that were now idle or flooded became prime hunting grounds. Other mining sites were sold off to private companies or investors such as Dan Gertler, pushing artisanal miners off the land, with little or no compensation. It's estimated that around this time over ninety percent of the Congo's cobalt was mined by hand. The Chinese quickly saw an opportunity to become buyers and many small traders moved to the Congo. China's economy was growing by double-digits, and the country was becoming a manufacturing centre for mobile phones. BYD, now the world's largest electric carmaker, started off making phone batteries in Shenzhen in the 1990s. The area became 'home to a rough capitalism reminiscent of that in the 1920s', author David Van Reybrouck wrote.[10] He visited one mine, Ruashi, which is now owned by a Chinese nickel giant Jinchuan, in 2006. 'I saw children clamber down into poorly shored-up wells of up to twelve metres (thirty-nine feet) deep. I saw a five-year-old boy covered in dust, wearing a "Plop the Gnome" T-shirt. If they were lucky, they received five dollars a sack.' Rory Carroll, a *Guardian* journalist who visited Kolwezi in 2006, captured the industrial desolation that forced miners to scavenge on old mining sites. He called it 'Stalingrad in the sun'. 'Potholed roads lead to ruined, rusted factories,' he wrote.[11] 'Trucks and bulldozers are lined up neatly, as if ready to roll, but the wheels are missing and the engines have cobwebs.' A US embassy cable wrote that miners worked in 'pre-Industrial Revolution conditions'.[12]

At first the Congo was not easy for Huayou, and the first visits to the country were conducted to do research, meeting with some of the earliest adventurers such as Billy Rautenbach, a controversial white Zimbabwean who had run Gécamines in 1998, as well as Groupe Bazano. 'We found that every process had a challenge behind it,' Lee recalled. But by the time Huayou realised how difficult it was they'd already invested so much money, he said.

Huayou ended up dominating the cobalt trade, buying up supply from other small Chinese traders who had also flocked to the area. The number of Chinese mining companies registered in the province of Haut-Katanga rose from fifteen in 2005 to over 100 by the end of 2013, according to the *Global Times*. 'CDM [Huayou's Congolese subsidiary] is like a funnel for all the Chinese,' Vital Kumungu from the charity Good Shepherd told me, spreading his arms wide to indicate the surrounding countryside as we drove along rutted roads in the hills near Kolwezi. He said he'd been to one place where 'if you see it, you'll cry', with pregnant women and children doing tasks while their parents worked. Soldiers guarded the mines so no one could get in, he said, indicating that some mining sites had protection from the government. Huayou became the 'king of ASM', Elisabeth Caesens, a researcher at the NGO Resource Matters, said.

Typically, ore from artisanal miners is sold to a trader, a *négociant*, who then takes it to local markets to sell, where it is weighed and examined – a process that many I spoke to claimed was rigged. According to Southern Africa Resource Watch, this rigging takes place 'with the full knowledge of the political and administrative authorities'.[13] A large buyer such as Huayou came to have considerable pricing power. While most Congolese miners earned between $1 and $2 a day, Huayou's earnings before interest and taxes were $267 million in 2011, which rose to $442 million by 2016. As Siddharth Kara, a Harvard expert on slave labour, has said: 'The most effective way for any business to increase profits is to minimise costs. For most business, the largest operating cost is labour.'[14] To expand its source of supply in 2008 Huayou bought three mining projects in the Congo and seven years later acquired the roughly thirteen-square-kilometre CMSK copper and cobalt concession from Gécamines for $52 million, giving the company ownership over the historic Luiswishi mine, which was one of the first deposits mined by the Belgians in 1900.

It was via Huayou that most of the cobalt mined by hand made its way to China, helping to fully integrate Congo into global supply chains and increasing China's hold over the country's resources. By 2019 China processed ninety percent of Congo's cobalt. Huayou became a key supplier of cobalt to batteries used in smartphones and computers, which China increasingly exported to the Western world. It supplied companies such as Sony, Nokia and Samsung Electronics, and Apple. Huayou quickly entered the supply chain for electric carmakers such as BYD and Volkswagen as well as battery makers LG Chem, Samsung SDI and CATL. It was fully integrated, from the mines in the Congo to the refining and production of battery materials in China. Huayou listed on the Shanghai Stock Exchange in 2015, and got backing from the China Africa-Development Fund, a fund set up by the state-owned China Development Bank, which owned ten percent of the shares. It became China's largest cobalt producer. A decade after Huayou took a gamble on the Congo over ninety percent of the minerals from around Kolwezi were shipped to China.[15] No one cared where the cobalt came from, how it was mined, or whether children were involved. As a result, consumers across the world were all indirectly complicit in the practice of child labour. The only thing that mattered was making the phones, and their supply chains, cheap. Designed in California, Made in China. But don't mention the Congo.

It would take the electric car and the work of a global NGO to properly shake the world out of its slumber.

*

In early 2016 the global human rights NGO Amnesty released a hard-hitting report that denounced Huayou's Congolese business for supporting child labour. The report, entitled 'This is what we die for', sent shockwaves throughout the industry and prompted a flood of media reports on the child labourers of Kolwezi.[16] Based on research at five mine sites a year earlier, the report detailed how children as young as seven scavenged for rocks containing cobalt in the 'discarded by-products of industrial mines', washing and sorting the ore before it was sold. Most miners, who spent long hours digging for cobalt, did not even have basic equipment such as gloves, work clothes or face-masks, they found. A number of children who were interviewed described how they worked twelve hours a day, carrying heavy loads, and earned between $1 and $2 a day. Even children who went to school worked as miners on the weekend and in school holidays as well as before and after school.

Kasulo was one of the areas Amnesty visited, which it found dotted with mining shafts, one of which led down through the floor of someone's room. Miners worked with mallets, chisels and head torches, climbing down into tunnels more than thirty metres deep in bare feet, then digging in horizontal tunnels that followed the rock seams. The digging was so disorganised and haphazard that 'at several points the tunnels meet those of neighbouring teams – the miners told researchers that they worry that at night when their mine is unguarded, their neighbours sometimes break into their tunnel to steal their ore'. Children were openly working on the surface of the mines at Kasulo, sorting and crushing cobalt ore, they added.

Amnesty said it was clear who the buyers of the cobalt were. It found that Huayou Cobalt's subsidiary, CDM, was the 'largest single buyer of cobalt that originates in the artisanal mines in and around Kolwezi'. It cited one Ministry of Mines official as saying 'CDM is the giant'. A Chinese businessman in charge of one of the buying houses at the buying market of Musompo said CDM is so big 'it is like America.' The report included a photo of a sign outside CDM's warehouse which said that it 'buys copper and cobalt products at a good price'. The researchers spotted Huayou's orange trucks being loaded with sacks of ore at Musompo and traced them to CDM's smelter in Lubumbashi. Amnesty was damning in its conclusions, saying Huayou was 'failing to respect human rights' and concluding that 'there is a high risk that Huayou Cobalt is buying (and subsequently selling) cobalt from artisanal mines in which children and adults work in hazardous conditions.'

The report also made clear that many of the big electronics companies had no idea where the cobalt they bought came from. The responses of some of the largest multinational corporations were eye-opening. Samsung SDI, one of the world's largest battery makers, said 'it is impossible for us to determine whether the cobalt supplied to Samsung SDI comes from DRC Katanga's mines'. Apple said only that it is 'currently evaluating dozens of different materials, including cobalt, in order to identify labour and environmental risks'. VW said that its 'internal system for sustainable supply chain management' had not identified human rights abuses in cobalt supply chains. 'It is a major paradox of the digital era that some of the world's richest, most innovative companies are able to market incredibly sophisticated devices without being required to show where they source raw materials for their components,' said Emmanuel Umpula, executive director of African Resources Watch, a local non-governmental organisation, who wrote the report with Amnesty.[17]

Amnesty's report landed on the desk of Huayou's headquarters in Tongxiang with an unpleasant thud. The company had been internationally shamed and faced the risk of losing its key customers. Apple immediately suspended Huayou from its supply chain pending an investigation, despite the fact that Apple employees were not allowed to visit the Congo to investigate the supply chain themselves due to company rules. There was no other option for Huayou but to clean up its act. The company appointed Lee as head of Corporate Social Responsibility, a new department and a new position, and formed a Corporate Social Responsibility Working Committee. It claimed that it wanted to be 'a world leader in ethical sourcing'. It undertook a mapping of its supply chain and conducted an audit, done by an American woman, Liz Muller, whose company was called Liz Muller LLC. It spruced up its website, and set Lee to travel the world to give talks. In a presentation Lee showed me in London in 2017 one slide was entitled, in bright red letters: 'Quite Shocked! Huayou never employs child.' Another bullet point said: 'Huayou didn't source directly but indirectly from cobalt ASM,' a get-out that was only the case due to the middlemen involved – the *négociants* and buying houses, of which Huayou was fully aware. The presentation said Huayou sourced between twenty and thirty percent of its cobalt from 'artisanal and small-scale mines', making it inconceivable that the company did not know what these sites were like. 'Huayou is a law-abiding and responsible enterprise and has a very good reputation in the DRC,' the presentation said. 'Pave roads, dig wells, develop agriculture, invest schools and engage in public welfare for the local communities.'

China's government also publicly pushed for change in the sector, creating a 'Responsible Cobalt Initiative' under the Chinese Chamber of Commerce for Metals, Minerals & Chemicals (CCCMC), which held a meeting in Beijing attended by Apple and Samsung.

Yet despite the rhetoric little changed on the ground and Huayou continued to buy from artisanal miners, creating bumper net profits of 1.9 billion RMB ($297 million) in 2017 as cobalt prices rose towards a ten-year high. While it claimed to have eliminated child labour, how could it be so sure? It could not monitor every site all the time – especially since all the cobalt got mixed in the local markets. Lee's presentation admitted as much, saying 'no matter how hard we trained the ASM supplier chain, we think it is very hard to control risks in the traditional ASM supply chain in the DRC'. The only option was to close the markets, Lee said. But it was not right to stop sourcing from artisanal miners altogether, he added.

Two years later on the train to Quzhou, Lee complained to me about other Chinese companies who had not been singled out by Amnesty and continued to buy from the Congo, taking advantage of the cheaper supply. Huayou could not create change in the Congo alone – what about Apple, what about the car companies, why did they not step in? Huayou could not spend a lot of money to change conditions on the ground. 'You can't keep on pointing fingers on one part of the supply chain and hoping it's going to get fixed,' Assheton Stewart Carter, a sustainability consultant, said. One of Huayou's Chinese rivals, Hanrui Cobalt, even went as far as to declare in its listing prospectus to the Shanghai Stock Exchange that it bought from traders in the Congo, without any details of how it avoided child labour. Lee's meaning was clear to me: if the other Chinese companies were still doing it, how could you claim to have fixed the problem?

*

The action by Amnesty echoed another movement involving the Congo that caught the world's attention more than a hundred years earlier just as the car industry was getting started. In the late nineteenth century, the Congo was ruled as a personal fiefdom by Belgium's King Leopold II, under the name of the Congo Free State. Ivory had been one of the country's main exports, and was turned into keys in pianos across Europe. But by the end of the century a new commodity was in demand: rubber. The invention of the inflatable tyre by John Dunlop in 1888 had prompted a bicycle craze that Leopold was only too keen to exploit. The ivory boom had come to an end a few years earlier, leaving Leopold almost bankrupt. He was only too happy that his colony was blessed with wild rubber vines that climbed into the trees of the equatorial forest. It was a prime opportunity for further revenues, which would all enrich Leopold personally, not the Congolese, or the Belgian taxpayer. The money went towards building monuments to Leopold in Belgium, including a colonial museum outside Brussels with a luxurious garden modelled on Versailles. To extract ever more rubber a system of forced labour was put in place, with each village given a quota to fulfil. 'Leopold had ridden the world's rubber boom like a man on a trapeze,' as historian Thomas Pakenham puts it.[18] 'Before the boom, the Congo's exports had consisted of a trickle of oil and ivory. By 1902, rubber sales had risen fifteen times in eighteen years, and constituted over eighty percent of exports, worth over forty-one million francs.' The rubber required manual labour to extract, meaning it was cheap – just as cobalt was many years later.

To maximise rubber output, each district in the Congo was given quotas and the much-feared Belgian military *Force Publique* were sent to villages to force young men to work. Men were punished if they didn't collect enough rubber, or if they refused to work. Stories and pictures emerged of severed hands, which had been brutally cut off as punishment. 'Severed hands became a way for sentries to justify collection shortfalls,' the historian Maya Jasanoff wrote.[19] 'If enough rubber hadn't been gathered, soldiers would kill natives merely to cut off their hands. Sometimes, to spare bullets, they just cut off the hands of the living instead.'

These events soon caught the attention of a young French-British man employed by a Liverpool shipping line which sent him regularly to Antwerp. Edmund Morel had been born in a suburb of Paris, the son of a French civil servant and an Englishwoman of Quaker descent. After his father died when he was four, his mother made a living as a music teacher in Paris and Morel was sent to boarding school in England, at Bedford Modern. When he was seventeen Morel got a job as a clerk at Elder Dempster in Liverpool. He also freelanced as a journalist by night, writing about West Africa. Initially unwilling to believe the stories of abuses in the Congo, by 1900 Morel made a startling discovery that would change the course of his life. He noticed that ships from the Congo were arriving full of ivory and rubber at Antwerp but returning only stocked with arms. The Congolese were clearly not being paid for their efforts. This discovery led Morel to unravel the brutality of Leopold's Congo: a personal fiefdom based on slave labour for the sole purpose of enriching the king, who never set foot in the country. Morel's 1904 book, *King Leopold's Rule in Africa*, and the follow-up, *Red Rubber*, exploded the myth. Morel led a global campaign with the Irishman Roger Casement, the British consul in the Congo, that put growing international pressure on Leopold and his Congo Free State. Newspapers and magazines showed pictures of burned villages and severed hands. Casement even gave Joseph Conrad, whose *Heart of Darkness* had been published in 1899, a copy of a pamphlet written by Morel.[20] 'It is as if the moral clock had been put back many hours,' Conrad wrote about the Congo. 'And the fact remains that ... seventy-five years or so after the abolition of the slave trade (because it was cruel) there exists in Africa a Congo State, created by the act of European Powers where ruthless, systematic cruelty towards the blacks is the basis of administration.'[21] Pressure from the Congo Reform Association founded by Casement and Morel was a key factor in forcing Leopold to hand over the Congo to the Belgian government in 1908.

Over a hundred years after Morel's campaign, the cobalt trade was not the result of a colonial power, but of the global, ruthlessly efficient market that only cared about cost. The metal was sent not to the docks of Antwerp but to the big ports along the Chinese eastern seaboard. It entered global supply chains just as the rubber had done, ending up in products used by all of us. During the process China took most of the value from the cobalt, which was exported from the Congo as crude cobalt hydroxide, an intermediate product that could not be directly used in industrial processes. Huayou's combination with LG Chem was an example of how China had solidified its position in the battery supply chain, turning raw materials from the Congo into high-value battery products. Huayou would not consider carrying out its smelting processes in the Congo, Lee told me, as it was too difficult to get hold of the reagents and other chemicals as well as reliable sources of energy. Besides, getting product out of the landlocked Congo was not easy, and the border crossings were known to be difficult. Glencore had had to spend $237 million building a plant to generate its own sulphuric acid to avoid trucking it in. There was also little chance of LG Chem choosing to build a plant in the green rolling hills around Kolwezi. As a result, Morel's question of Leopold's time could still be asked: the Congolese sent their cobalt, but what did they get in return?

*

In 2018 at the annual DRC Mining Week a white Tesla Model X was brought to the city of Lubumbashi, and exhibited in the garden of the city's best hotel to show to the crowds. It was a jarring image – few people in the Congo would be driving an electric car any time soon and there were no superchargers in the country. The electric car represented a new front in the persistent inequality of globalisation and trade. Benjamin Sovacool, an academic at

Sussex University who had written a study on artisanal miners of Kolwezi, called it the 'decarbonisation divide'.[22] As demand for electric cars rose, miners in Kolwezi responded by digging more cobalt out of the ground, in basic conditions. They responded to global prices as much as anyone along the global supply chain, yet they received the least value. Money from the trade did not help the Congo move up the value chain or even become prosperous. In 2018 it was estimated that seventy-three percent of the population, around sixty million people, lived on less than $1.90 a day.

That was a problem for the electric car industry, since the EV was supposed to be green and beneficial to society. Buyers of electric cars were interested in 'virtue-signalling', or proving their green credentials. 'Emotionally an EV is supposed to be a good deed – you're buying an EV thinking you are saving the planet – the last thing you want to hear is that the car is not clean,' Nicholas Garrett, chief executive of RCS Global, a supply chain tracking company, told me.

Yet while the Amnesty report had highlighted the plight of the miners as well as the power of global NGOs, Amnesty soon discovered the unintended consequences of their probity – especially with the car companies. The images of children labouring in the mines of the Congo had been effective in raising attention and getting journalists interested in travelling to Kolwezi. A Sky News report on two child miners, Richard and Dorsen, was viewed by millions and helped raise money to get them moved to shelter and into protection. But the media attention only increased the risks to car and battery companies of buying cobalt from the Congo. They realised that images of children digging up cobalt that ended up in an electric car could ruin their reputations overnight – and that such images were always only one click away on social media. China's state-owned newspaper *Global Times* even went so far as to say that the Sky News report had been staged 'as part of a Western conspiracy against Chinese companies'.[23]

One quick and simple solution was to avoid buying from the Congo altogether. 'No one wants to touch ASM [artisanal mining] with a bargepole,' one head of procurement at a carmaker said. In April 2019 at an OECD meeting in Paris on responsible sourcing Andreas Wendt, who was responsible for BMW's procurement and supply chain, proudly stood up and said the car company would not be sourcing from the Congo at all, but buying from Australia and Morocco instead. His message was simple: consumers could feel assured that the car they were driving did not contain a piece of metal that had travelled from the dusty mines of the Congo into their garage. It was comforting to ignore the Congo, to wish it out of existence. Monique Lempers, from the sustainable phone maker Fairphone, said the Amnesty report had made companies adopt a 'risk-averse approach', to ensure there was no child mining of cobalt, rather than to engage with the issue. 'There was a lot of resistance from Western companies to engage in ASM [artisanal] cobalt, whereas the fact you are sourcing from a mine that is not yet perfect but showing improvement is far better than saying it doesn't exist when we know the problem is out there and flowing somewhere in the supply chain.'

The other option was to buy from Ivan Glasenberg's Glencore, whose cobalt was dug up behind heavily armed fences, cut off from the chaotic artisanal mining sites such as Kasulo. In April 2018 the miner publicly warned that sourcing cobalt from small-scale miners in the Congo increased the risk of child labour. 'As a major producer and marketer of cobalt, we support efforts to establish greater transparency in the value chain, and address the endemic poverty in this region that is the underlying cause of artisanal mining,' Glencore said. 'We do not support ASM [artisanal mining], nor process or purchase any material derived from ASM in the DRC.'[24]

The allegations of child labour in the Congo created a key competitive advantage for the big industrial mining companies. In late 2017 at a conference in New York the World Economic Forum launched the Global Battery Alliance, which included miners such as Glencore and Kazakhstan's ERG, which was led by former BCG management consultant Benedikt Sobotka. ERG's predecessor company had been delisted from the London Stock Exchange amid an investigation by the UK's anti-fraud agency, but now Sobotka was a regular at WEF meetings and at Davos. The Global Battery Alliance was launched with the aim of creating an 'ethical and sustainable global supply chain for ... lithium-ion batteries'. In its press release announcing the launch, Sobotka was keen to state that its 'large industrialised complexes' produced cobalt free of child labour. 'Unfortunately, there is almost a 100% chance that your smartphone or electric vehicle contains cobalt that comes from child workers in artisanal mines. Although creating new ethical energy sources will help, we all need to do whatever we can to put an end to child labour,' Sobotka said.[25]

The big mining companies wanted to conflate their own supply with the concepts of 'ethical' and 'sustainable'. They asked the LME, the world's centre for metals trading, to remove metal that might have come from artisanal sources in the Congo from its warehouses. Matthew Chamberlain, then the fresh-faced thirty-four-year-old chief executive of the LME, found himself at the centre of the polarised debate. Chamberlain, a computer science graduate from Cambridge, had been an adviser on the takeover of the 138-year-old exchange a few years earlier by Hong Kong's stock exchange operator Hong Kong Exchanges and Clearing. The miners complained to the LME that Chinese cobalt metal in its warehouses was of unknown origin from the Congo, which was deterring people from trading the LME's cobalt contract, preventing a liquid global price for cobalt. The metal came from the small Chinese producer Yantai Cash, which they said should be delisted.

Chamberlain moved to act. In October 2018 the LME said suppliers of cobalt needed to comply with responsible sourcing standards by the end of 2020 or they would be removed from the exchange. But whose 'responsible sourcing' standards? The move drew instant criticism from fourteen NGOs including Amnesty, who pointed out the corruption allegations involving Glencore. The LME's move was a 'greenwashing mechanism', they said. 'Corruption is a supply chain risk that has been long-overlooked by companies engaged in responsible sourcing, but which carries heavy consequences.'[26] The LME was soon forced to backtrack, saying it would introduce responsible sourcing for all metals and all suppliers.

For carmakers, Glencore quickly became a preferred supplier of cobalt, however. While the company was under investigation on corruption charges by both the UK Serious Fraud Office and the US DoJ and was paying royalties to Israeli billionaire Dan Gertler, who was under US sanctions, it was preferable to the risk of child labour. In late 2019 Glencore signed agreements to supply cobalt to Korean battery makers SK Innovation and Samsung SDI – for batteries that ended up in electric cars. The company's willingness to enter the Congo was beginning to pay off. But to others, Glencore's links to Gertler remained troubling. Elisabeth Caesens of Resource Matters called these supply deals 'See No Evil, Speak No Evil'. 'For every $100 spent on cobalt from Glencore, more than $2 is owed to a company under US sanctions,' she said. 'Companies that want to make sure they're not linked to illicit deals need to look at the Glencore-Gertler connection first. Put simply, if they can't show they have asked tough questions of Glencore about payments to Dan Gertler and got satisfactory answers, they can hardly claim to be clean.'[27] One industry executive said

Amnesty's report had backfired. 'The NGOs created their own problem,' he told me. 'They realized that opposition to artisanal mining would hand a big cheque to Glencore.'

*

The Congolese government was growing increasingly concerned, however, that the reputation of artisanal miners risked hastening carmakers' efforts to remove cobalt from their lithium batteries, denying them an opportunity to benefit from the coming boom. In April 2018 Congolese ministers travelled to the annual OECD meeting on 'responsible mineral supply chains' in Paris to reassure attendees they were tackling child labour in the cobalt trade. A year later Albert Yuma, the former head of the state mine Gécamines, launched into an attack on the artisanal mining sector at the DRC Mining Week in Lubumbashi, that was mentioned earlier. Yuma took the stage after attendees observed a minute's silence for those who had died in artisanal mining. He described the chaos of the system, saying individual miners had caused the country to lose millions of dollars in revenues by selling to foreign middlemen at prices lower than on the international market. In 2018 prices had surged to a ten-year high of $40 a pound, leading to a flood of supply from artisanal miners in and around Kolwezi. Miners could earn good money digging out cobalt, but not nearly as much as the Chinese buyers. Yuma realised that the Chinese had captured all the benefits under the nose of the Congolese state, which had been powerless to stop them due to corruption and the involvement of local elites. As Southern Africa Resources Watch put it, the Congo had given 'carte blanche ... knowingly or negligently, to foreigners dominating the artisanal sector, which is supposed to be under the exclusive control of the Congolese with a view to accelerating local development'. To make matters worse, the resulting surge in cobalt quickly led to a price crash the following year, when prices slumped to $14 a pound – just as the Congolese government was looking to earn more royalties from the mining sector. Those who were digging for cobalt quickly found out they were not in line for the riches they had expected. They had unwittingly crashed the price of their own product. With few other opportunities in Kolwezi that could generate the money earned from digging for cobalt, thousands of miners faced ruin. There was a deeper fear too: if carmakers considered cobalt a 'blood mineral', something too dangerous to touch, then the Congo risked losing out on its God-given monopoly. The boom would be over before it had even begun. The solution was to impose some sort of control on the industry.

*

Driving into Kolwezi there is little doubt where all the country's metal goes: the road into the town is surrounded on both sides with buying stations with names like Boss Wu or '888' (lucky numbers in China) that will buy from whoever brings cobalt and copper to their door, no questions asked. The symbols for the metals 'Cu, Co' are painted on the walls. The dusty city is dotted with Chinese restaurants and a Chinese casino where gamblers can try their luck in the poorest country in the world. Like miners the world over, those who make money from cobalt spend it quickly.

Facing the risk of a full-scale physical collapse of the Kasulo site, in 2017 the authorities in Lualaba province evacuated all six hundred households from the site. Huayou paid for the forced relocation in return for the right to continue to buy all the cobalt from the area, thereby avoiding buying from the local markets where all the cobalt was mixed. Residents were offered new houses from Huayou or cash, and most of them took cash, Lee said. After the forty-hectare area was cleared of houses a perimeter wall was erected around the site and a security gate introduced to check miners on arrival. The area would now become what Lee called a 'model mine', a way to make artisanal mining safer in the hope that it could enter the electric car supply chain. RCS Global, a Berlin-based chain audit company, would monitor the area, supplying data to a consortium that included Ford, Volkswagen, LG Chem and Huayou. (Chinese-owned Volvo also later joined the network. Glencore and Tesla would join a group called the Fair Cobalt Alliance, which also worked with RCS Global.) The theory was that once they had data, the companies could at least be confident that conditions were improving, even if they weren't perfect. Eventually they hoped for the emergence of an artisanal standard for cobalt, similar to those for Fairtrade coffee, which actually gave coffee farmers a premium over market prices. (Most artisanal cobalt was sold at a discount to market prices.)

When I visited, I was taken in the smartest, newest Toyota Land Cruiser I've ever seen – Huayou's official car – along the bad roads on the short drive outside Kolwezi. On the road we passed beat-up local taxis, as well as the local overloaded mini-buses. We passed a roundabout in town with an iron statue of a miner crouched in work, which had been paid for by one of the earliest adventurers, the Lebanese businessman behind Groupe Bazano. We drove through the middle of the bustling village of Kasulo to arrive at the high metal gates of the mining site. Inside, there was no sign of the families that previously lived there, or the chaotic tunnels. The land had been cleared, and overburden removed, so miners worked under orange tarpaulins at the bottom of one large open pit – reducing the amount of earth they had to remove. About 600 miners worked on the site, down from around 5,000 the previous year, though this was partly due to the fall in cobalt prices. A Chinese security guard from Changsha who accompanied us pointed proudly to a sign that said no children or alcohol were allowed. Wearing a loose-fitting black flat cap and faded black jacket, he wouldn't have been out of place in a new

Chinese apartment complex. He also pointed out bathrooms and a small clinic built by Huayou.

The miners are organised into co-operatives which take a cut of the number of bags sold by the workers in return for assistance such as covering medical bills, helping family members in the case of death and representing the miners at political meetings. After digging the cobalt, the miners take their sacks to be crushed, weighed and graded in on-site depots, after which the material is authenticated and sold to local traders and then to Huayou. A hand-held machine with Chinese language software that looks like a barcode reader scans the ore and reveals the percentage of metals inside. The details are then written down on Chinese receipts, the *fapiaos* found all across China. The trucks heading out of the site are also checked and sealed to make sure they are not tampered with before they arrive at Huayou's subsidiary, CDM. A smartphone app is used to report accidents or deaths, or forced labour. These reports go straight to a central data point – with immediate alerts sent to Huayou, said Robert Bitumba, a genial broad-shouldered Congolese who was born in Kolwezi and studied for an MBA in South Africa. He helps monitor the site for RCS Global. The theory is that if conditions on site deteriorate, RCS can directly push Huayou to help improve them, since the data can be shared with its customers such as carmakers. That's different to other Chinese companies in the DRC who fail to do even 'basic due diligence', Bitumba told me. 'People were digging without any control,' he said. 'It was a very dangerous situation so the government had to deal with that.'

There are no children on the site now, Bitumba told me, and fewer deaths. In the three months after the project began operating in July 2018, there were three deaths, but since then there had been none, RCS told me. There were five recorded instances of child labour between July and September, but this had also since fallen to zero. Lee told me Huayou was looking at scaling up the model to other sites in the Congo. If it doesn't, he said, tainted cobalt will continue to be sold on the open market, putting everybody in the supply chain at risk. 'If the people were still living in Kasulo then there is no way to get rid of the child labour risk,' he said. 'We can copy the model and try to assess it.' Still, when I asked Bitumba about safety equipment on the site, he told me it was still sitting in a container and hadn't been given out yet.

While it is supported by the Lualaba government, the Kasulo project still has its critics. The relocation of the villagers triggered protests and has been opposed by some organisations in Kolwezi. Questions have also been raised about who controls the co-operatives and how connected they are to the government. The Good Shepherd charity, which helps children leave mine sites by providing free schooling, decided to end its engagement with Huayou over concerns about the relocation of the residents and their compensation packages. Emmanuel Umpula of African Resources Watch told me local miners should be able to sell their cobalt to whoever they want to rather than being forced to sell all their metal to one company, Huayou. 'We need zones where the miners should be in co-operatives created by themselves and where they can work and sell their minerals to buyers,' he said. 'We need them to have the possibility to say "this is not a good price" and go elsewhere to sell their minerals.'

Distrust remained high when I visited. Some miners recently broke in and damaged the platform that weighs cobalt trucks, angry at the lower prices for their metal. And no sooner had the Kasulo project begun than a new site opened up outside its walls. The controlled site had to expand to include Kasulo II – a ten-hectare extension, with further relocations. Kasulo III is under consideration. Bitumba says that in the long run the country must provide alternatives to mining, otherwise people from elsewhere in the Congo will continue to flock to the Kolwezi mines. 'Almost everywhere [in Kolwezi] there are minerals,' he said. 'For me it's a good thing to have a fence because in that area you then have control. But now you need to put in incentives for all the artisanal miners to go inside that fenced area.' In the long run the economy needed alternatives to mining, Bitumba said on our journey back to Kolwezi in the afternoon heat. 'You have to have good economic growth in the country and other opportunities. Also to encourage people to do agriculture and things like that.'

Most critically, while conditions at Kasulo had improved, it had not yet reached a standard where car companies and other international buyers were willing to buy cobalt from the area. In the summer of 2020 in the face of pressure from its clients, Huayou said it would no longer buy from Kasulo or other artisanal areas, though many critics doubted whether that would actually happen. Until cobalt from such sites could safely enter the supply chain, there would be little long-term incentive for Huayou and others to stick with projects like Kasulo.

The challenges at Kasulo – let alone other sites in the area – were daunting. Fully eradicating child labour in the region would require improving the economy and job prospects, not something that could be solved by one company alone. And even if miners were better controlled at Kasulo the health impacts of artisanal mining were still present. The controlled sites give the illusion of safety, according to Professor Nemery. It will take a lot of money and effort to improve the lives of artisanal miners, he told me. 'We don't know about long-term health effects,' he explained. 'The workers still work in the same poor circumstances apart from some symbolic gestures. To really have prevention of accidents and disease you need much, much, more. Health and safety at work is much more than giving personal protection equipment. We have to be careful not to impose simplistic solutions that make things worse.' But Nemery admitted that banning artisanal mining was not an option either. 'So, unfortunately, sustainable cobalt mining in the DRC is still a utopia,' he said.

Just as carmakers began to engage with the problems and complexities of cobalt mining in the Congo, however, another problem reared its head that stood in the way of their ambition to go fully electric: nickel.

10

Dirty Nickel

'Please mine more nickel ... Tesla will give you a giant contract for a long period of time if you mine nickel efficiently and in an environmentally sensitive way ... Hopefully this message goes out to all mining companies. Please get nickel!'

Elon Musk, Tesla CEO[1]

In the summer of 2019, Swiss mining consultant Alex Mojon was invited to the island of Papua New Guinea (PNG) in the South Pacific to carry out an environmental assessment of a Chinese-run nickel and cobalt mining project that had been disposing of waste into a local bay. The mine's processing plant, an area of circular tanks and smoke-emitting chimneys, had been carved out of the forest, right next to the water on PNG's north coast. The nickel was mined further inland in the Kurukumbari mountains, and then sent through a 134-kilometre pipeline to the refinery. The rock dug up inland contained less than one percent nickel, meaning there was a huge amount of waste to be disposed of and forests had to be cleared to mine enough of it. Normally mining companies built large dams on land to store mining waste, but the Ramu mine disposed of its waste directly into the sea, at a depth of 150 metres – a method known as deep sea tailings displacement. The theory was that the waste, being heavier, would slip down to the ocean floor at around 1,500 metres where it would rest undisturbed.

A few months after his first visit a series of pipeline leaks at the refinery turned the local sea red. There were reports of dead fish and 'people developing complications after allegedly eating contaminated fish and swimming in the sea'.[2] The local Madang provincial government called it the worst disaster in Papua New Guinea's history and put in place a fishing ban, cutting off one of the main sources of livelihood for locals.

Mojon returned to PNG in September for a second visit. Accompanied by soldiers Mojon, a short and stocky man in a khaki waistcoat, took food and water samples from local villages to take back to a laboratory in Munich for testing. One woman brought out her child who she said she believed may have been deformed by the activities of the Ramu mine. In one district he saw a dead dolphin that had been washed up onto the beach, and took a sample from it. On a third trip in October Mojon discovered algae-like brown-red filaments floating around three kilometres out to sea and ten kilometres north-west of the plant. Local fishermen told him they had never seen such vegetation before.

Mojon, who had a long career as a mining and petroleum geologist, was shocked by the amount of waste the Ramu mine was routinely dumping into the ocean. While the waste itself was not that dangerous, its sheer volume could smother the seabed or marine ecosystems if it resurfaced. A total of 77.6 tonnes per hour of mine waste was going into Basamuk Bay, or 680,000 tonnes a year, 'twenty-four hours a day 365 days a year – giant volumes!' Mojon recalled. He found that fine particles in the mine waste remained in suspension in the sea water and were carried by currents over a much larger area, including to beaches of nearby islands. The waste did not sink to the bottom of the ocean as Ramu had planned. Instead, he believed it contaminated local marine fauna and island coast-lines. It was almost definite, he said, that pumping mine waste to 150 metres would lead to the deaths of local fauna and flora. Testing in Germany of samples from 'agricultural soils, beach sand, water from river and sea, including drinking water, and food such as fishes, crabs and Taro potatoes' all showed very high concentrations of heavy metals above allowable limits, he said. Overall Mojon concluded that the mine waste would 'have a drastic, non-reversible impact on the eco-systems, marine life and humankind health in totality'.

The Ramu mine had been bought in 2005 by a Chinese state-owned company, China Metallurgical Group Corporation, which was now a subsidiary of China's largest state-owned metals giant, Minmetals. The company had bought the Ramu nickel project at the height of China's efforts to 'go global'. But China had entered a country where mining had left a terrible legacy. In 1988 an uprising sparked by a copper mine in the autonomous region of Bougainville led to a civil war,[3] with the death of an estimated 15,000 people. The mine, which had been owned by mining giant Rio Tinto, had exacerbated local tensions and also despoiled the environment by dumping waste into the Jaba River. 'For PNG, the promise of mining-led development remains elusive to many communities and they are invariably left with significant social and environmental legacies which will last for decades to centuries (e.g. mine waste impacts on water resources),' one study concluded.[4] Years later the pollution from the mine was still causing problems. Matthew Allen, a professor at the University of the South Pacific, found in 2015 that waste rock downstream from the mine was still at least a kilometre wide at its greatest point. 'Every new rainfall brought more tailings downstream and changed the course of the waterways, making life especially challenging for the hundreds of people who eke out a precarious existence panning the tailings for remnants of gold,' he wrote.[5]

Mojon revealed his findings at a press conference held in the capital Port Moresby with Governor Peter Yama in November

2019. He proposed taking immediate measurements to assess the damage from the spills as well as a programme to restore the contaminated area. He also proposed alternatives to deep sea disposal of mining waste. But nothing happened. Mojon was not allowed to finish his environmental assessment of the Ramu mine and left the following month. Ramu said it would only cooperate with the central government's enquiry into the spill – which ascertained that the waters were safe. In Parliament minister Geoffrey Kama had already announced the results: 'Though the discolouration caused some fear amongst the locals, everyone must understand that "the immense volume of the sea waters is an excellent buffering solution for the acidic slurry" – hence it immediately diluted and dissipated the spilled slurry,' he said.[6] Mojon left PNG frustrated and with a sense that China's power over the small island country had allowed Ramu to get away with dumping waste and destroying the environment. On returning to Switzerland, he reflected that society spent so much time obsessing over eating organic food but had no issue with dumping thousands of tonnes of waste into the ocean.

Investors in Ramu, however, voted with their feet, with Storebrand, Norway's largest private asset manager, divesting from the Hong Kong-listed Metallurgical Corporation of China (MCC), citing unacceptable levels of environmental damage. 'Dumping of mining tailings directly into marine environments is a controversial practice internationally. Marine ecosystems are crucial to a healthy planet and must be protected,' said Bård Bringedal, chief investment officer at Storebrand Asset Management.[7] More than 5,000 villagers and the provincial government in PNG also mounted a legal challenge against the company the following year, in February 2020. They demanded a total of $5.2 billion in compensation for damages and for the company to stop dumping mine waste into waters.[8] None of this, however, stopped Ramu from operating and delivering nickel to China.

The Ramu case alarmed many because the spills happened just as a host of Chinese companies were looking to start nickel projects next door in Indonesia, which would also generate large amounts of mining waste. The projects were intended to serve the electric car battery market. As carmakers reduced the amount of cobalt in their batteries, they increased the amount of nickel. Using more nickel increased the amount of energy the batteries could store – allowing cars to go further on one charge. Tesla had always used more nickel than other carmakers, which had been critical to assuaging buyers' concerns about 'range anxiety'. Nickel would also be critical for Tesla's efforts to launch an electric pick-up truck, the Cybertruck, as well as a full-sized truck, the Semi. As Musk pointed out, energy-dense batteries (those that stored more energy per kilogram of weight) were critical for trucks since 'every unit you add in a battery pack, you have to subtract in cargo'.[9]

Yet when the car companies turned their attention to the nickel market it was not a pleasant vista. While nickel is mined in Russia, Australia and Canada, all of the new growth supply was set to come from Indonesia, where Chinese companies had invested heavily in smelters and stainless-steel plants. The government had also recently banned the export of raw nickel ore.

By the summer of 2020 Musk was becoming concerned about the availability of nickel. In August he said that Tesla would give a 'giant contract' to any company that could mine nickel in an 'environmentally sensitive way'. At the company's Battery Day presentation in September, he repeated his plea: 'In order to scale, we really need to make sure that we're not constrained by total nickel availability,' he said. 'I actually spoke with the CEOs of the biggest mining companies in the world and said, "please make more nickel, it's very important."'[10] Musk had held calls with the chief executives of the three largest nickel mining companies, Glencore, Australia's BHP and Brazil's Vale. But these companies had few plans to expand nickel production just for Tesla. They would not be able to meet Tesla's goals of advancing sustainable energy globally and becoming a true global car manufacturer. For that the electric car industry would have to turn yet again to China. As Tesla had found with cobalt, nickel came with a whole host of sustainability issues that would not only threaten the EV's image as a clean car but also carmakers' ability to scale up and electrify the planet.

*

Nickel had been admired for its silvery lustre for thousands of years, though it was only isolated as an element in the eighteenth century. A zinc-nickel alloy called 'baitong', or white copper, was used in China as far back as the fourth century CE and was exported to Europe from the southern port city of Guangzhou during the seventeenth to nineteenth centuries.[11] In Europe it was called by its Cantonese name 'paktung' and used for decorative items such as candlesticks and candle snuffers. But Europeans struggled to work out what paktung was made of and produce it themselves. China had concealed the method of making paktung, which involved smelting copper with a nickel ore, pentlandite, six or seven times.[12] It wasn't until nickel was first isolated in Sweden by Axel Fredrik Cronstedt in 1751 that Europeans came to understand the metal. The name was short for the German '*Kupfernickel*', which means 'goblin's copper'. The first breakthrough in terms of finding an industrial use for nickel, however, came over 150 years later with the invention of stainless steel, at a time when Europe, the pre-eminent economic centre of the world, was posed on the verge of self-destruction. In 1913 Harry Brearley, a metallurgist in Sheffield, accidentally discovered that adding a mix of carbon and chromium to steel created a new type of metal, one with a self-created protective layer that prevented the metal from rusting. The protective layer of chromium oxide also prevents us from tasting the metal when we use it for knives and forks. Nickel was soon also added along with chromium as a key ingredient for the metal. Stainless steel turbocharged nickel demand – nearly eighty percent of all the nickel mined was extracted over the past three decades, most of it from Canada. Stainless steel is now everywhere in our lives – from our kitchen sink to our cutlery. But few of us have ever thought about where it comes from or how the process impacts the environment. There are very few people willing to protest or even think about the metals in their cutlery. Such utilitarian objects have long been taken for granted, even though they too exact their toll on the earth.

The battery industry had fixated on nickel availability for decades. In 1901 Thomas Edison travelled to Canada to look for nickel, discovering a deposit in the Falconbridge area of Sudbury, Ontario, which would go on to become the largest nickel-producing area in the world. Edison, who at that time had already invented the light bulb and the phonograph, had seen a block of Sudbury ore at the Pan-American Exposition in Buffalo, New York. Along with his wife and brother-in-law, Edison used an electrical dip needle he had invented, called a 'magnetometer', to search for nickel in the Falconbridge area. In 1902 and 1903 the elderly Edison tried to sink a shaft numerous times in an area where he had detected a magnetic anomaly but was halted by a layer of quicksand. He gave up and returned to New Jersey. A couple of years later the nickel ore was found, a mere 4.5 metres lower than the bottom of Edison's original shaft.[13] The company that eventually developed the area, Falconbridge Nickel Mines, sunk its first shaft in 1928 and went on to become one of the largest nickel companies in the world, which was eventually acquired by Xstrata in 2006, which in turn was taken over by Glencore in 2013. Edison to this day, however, is credited in Falconbridge as being the first to find its nickel ore.

Nickel is found in two types of deposits – known as sulphides and laterites. The Canadian deposits were sulphides, which were found far from the equator in Russia and South Africa, and had been formed out of crystallised magma. They were generally easier to process into nickel for batteries due to the ability to form a concentrate prior to smelting, which vastly improved efficiency. The more abundant form of nickel deposits, however, accounting for three-quarters of the world's reserves, were the laterites, which were found mostly in the tropics, in Indonesia, the Philippines and New Caledonia, and were developed through weathering of base rock. (Australia had the best of both worlds with both laterite- and sulphide-type deposits, another example of its geological luck.) Mining the rusty-red-coloured laterite deposits, however, required much more energy, in order to separate the nickel from its tight bonds with iron. Most of that energy came from coal.

*

In May 2019 metal traders in London noticed something strange going on in the nickel market. Someone was buying up huge stocks of nickel on the LME, pushing up the price. It looked like an attempt to corner the market – a practice that had occurred from time to time but most famously in 1995 when a Japanese trader for Sumitomo, Yasuo Hamanaka, became known as 'Mr Copper' for buying up all the copper stocks on the exchange. The tactic, known as a 'short squeeze', is intended to give the false impression that demand for the metal is especially strong, prompting others to buy into the rally. It seemed to be happening again in nickel, since the real physical market for the metal was actually very weak, especially in China – but who was the buyer? The purchases were taking place via a US bank, according to the rumours, which pointed to JP Morgan. By October the stocks of nickel metal on the LME, which is a physically backed market, fell to their lowest levels in seven years, down by a third in just two weeks. Whoever was buying was succeeding – the nickel price was closing in on its highest level in five years.

The buying coincided with news that Indonesia, the world's biggest producer,* could bring forward a proposed ban on exports of raw nickel by two years in order to stimulate domestic processing and smelting – which was helping to push prices up even further. In August Indonesian officials confirmed the ban would begin in January 2020, surprising many in the market. The measure would immediately cut off shipments of nickel from Indonesia, leaving stainless steel producers stranded from one of their key raw materials. It was also a moment that the battery industry was watching closely. With almost every electric car in the world using nickel, a move by the biggest producer to cut off supply was reminiscent of the antics of Saudi Arabia and OPEC (the Organization of the Petroleum Exporting Countries) in the 1970s.

In the close-knit world of London's metal traders (a group of mostly men, who every October came together during LME Week to frequent the most expensive bars and clubs in London) people pointed at a Chinese company with huge operations in Indonesia: Tsingshan. A company that few people had heard of, Tsingshan was a member of the Fortune 500 and had revenues of over $30 billion a year – roughly what Tesla made in a year. In the space of a decade the company had decimated the global competition. Republican politician and former presidential contender Mitt Romney said China had 'achieved a breath-taking capture of the global steel market' by moving production to Indonesia. 'Indonesia just happens to be the largest producer of the world's nickel,' he wrote in an op-ed for the *Washington Post*. 'Suddenly,

---

* Indonesia produced 771,000 tonnes of nickel in 2020, twice that of the Philippines and around a third of global output. This was set to rise to 2.5 million tonnes of nickel by 2030, due to Chinese investment.

Indonesia has agreed to shut off nickel exports to any of China's foreign competitors. Another near-monopoly is born, thanks to anti-competitive tactics.'[14]

Indonesia's export ban meant that anyone who wanted nickel from the world's biggest producer needed to have set up shop in Indonesia – which was just what Tsingshan had done over the past few years.

It was estimated that Tsingshan bought around $1.4 billion worth of stocks of nickel on the LME during 2019 – at the same time as Indonesia announced its ore ban. Artificially boosting the nickel price by buying stocks of the metal made Tsingshan's rivals in Europe and China less cost-competitive. Finnish stainless steel producer Outokumpu said in October of 2019 that earnings had fallen by fifty percent in the third quarter from the previous quarter due to higher nickel prices and a €31 million loss due to hedging on nickel prices. The price of nickel had increased just as demand for stainless steel in Europe was falling – squeezing European producers at both ends. The Indonesian ban immediately led to trade tensions and accusations. In November of 2019 the European Commission wrote to the World Trade Organization to protest against Indonesia's export ban, saying it would 'unfairly limit access of EU producers to raw materials for steel production, notably nickel'.[15]

A month before news of the ban, Tsingshan's enigmatic founder and chairman, the slim and smooth Xiang Guangda, had met with Indonesia's President Jokowi at the Presidential Palace, along with the head of Huayou Cobalt, Chen Xuehua, and Li Changdong from China's giant battery maker CATL. Present at the meeting was Jokowi's powerful right-hand man, the former general and long-time business partner, Luhut Binsar Pandjaitan. Jokowi had first met Luhut in 2007 when they had set up a wood processing venture together.[16] Luhut had 'wide-ranging formal and informal powers' according to former journalist Ben Bland, most importantly over large investment decisions. The Chinese delegation – representing a large chunk of the global battery supply chain in one room – were all dressed in Indonesian shirts. Xiang praised 'Indonesia's rich nickel resources' and announced plans to increase investment in the country's Morowali Industrial Park from $8 billion to more than $15 billion. The article also intriguingly noted that Xiang 'offered several policy suggestions and proposals to President Joko on further optimising the investment environment in Indonesia'.[17] Could one of those suggestions have been to bring forward the ore ban just as Tsingshan was buying up supplies of the metal? 'The Chinese ... lobbied very intensely to completely ban exports,' Jim Lennon, a nickel expert at the Australian bank Macquarie, told me. The ban would make sure the nickel ore in Indonesia was kept available for these Chinese companies to use. Jokowi praised Tsingshan's investments in the country, saying: 'it is highly commendable that you are not only using Indonesia's resources to produce semi-finished products but also to produce finished products and drive the development of the downstream industries'.[18]

*

Xiang Guangda lived and worked from a four-floor converted old house in Shanghai's French Concession, where he had separate kitchens and chefs for European and Chinese food. Always casually dressed without a tie, he had a fleet of luxury cars including Mercedes-Benzes, Hummers and Bentleys, though he could not drive, according to a person who knew him well. 'He's not ostentatious personally but he likes to show off that he's got nice things,' they told me. 'He's extremely quiet – he's small of stature and doesn't say much. He listens very carefully and when he does speak it shows a towering intellect.' Another person who travelled with him for two days said he dressed modestly and carried only a small suitcase, wearing the same clothes. But in meetings he showed 'great business acumen'. Xiang was also ruthlessly focused and ambitious.

Tsingshan had its origins in the explosion of private enterprise in the city of Wenzhou, in China's eastern Zhejiang province, in the 1980s, following reforms brought in by Deng Xiaoping to loosen the state's grip on the economy. Xiang had been born into an ordinary working family in 1958 and worked at a state-owned marine fisheries company doing machine repairs. He was quickly promoted to become director of the workshop. As Wenzhou's private economy began to boom Xiang and his relative Zhang Jimin decided to leave the security of the state sector and founded a company to make car doors and windows in 1988. They eventually set up a stainless steel company in 1992, which was China's first private steel producer.[19] The company expanded production into Fujian and Guangdong provinces. While Xiang was not a Communist Party member, he retained deep links to these provincial governments. (A Guangdong state-owned enterprise would become a twenty-five percent shareholder in one of Tsingshan's key Indonesian subsidiaries.) In the mid-2000s Tsingshan helped develop a breakthrough way of making stainless steel by producing molten nickel pig iron and hot charging this directly into a stainless steel plant, which significantly lowered costs.

But by the time of the global financial crisis Tsingshan, and the rest of China's steel industry, was facing a shortage of nickel. 'Sixty to seventy percent of our stainless steel is nickel, but who produced this nickel? It was foreigners who produced it, China had none of its own production,' Xiang recalled.[20] 'But, if we wanted to develop stainless steel we had to resolve the problem of a shortage of nickel. Therefore, ten or so years ago, I believed we needed to develop in the area of nickel. If we didn't do this, this industry would have found it hard to continue.' It was a similar situation China had faced in cobalt before its companies went to the Congo, and lithium, before Ganfeng and Tianqi Lithium went to Australia and Chile. In 2009 Xiang made a similar move and decided to go to the world's biggest producer, Indonesia, to invest in nickel mining. 'The decision was forced upon them,' due to the shortage of nickel, the person who knew Xiang well told me. It would later pay off handsomely.

In late 2013 President Xi Jinping stood up behind a lectern at Nazarbayev University in Astana, the capital of Kazakhstan, to launch his grand Silk Road strategy to knit China closer to countries in Central Asia via infrastructure investment. A month later Xi made a speech to the Indonesian Parliament, declaring the launch of a twenty-first-century 'Maritime Silk Road' going from China across South-East Asia to East Africa. Both projects would later be amalgamated into the 'Belt and Road Initiative' (BRI), one of Xi's signature foreign policies. Today the Belt and Road project counts some 139 countries as supporters, among them those with rich resources of battery materials such as Chile and the Democratic Republic of the Congo. The BRI's scale and ambition is also limitless: it is prepared to exploit the melting Arctic and has launched satellites into outer space.

Following his speech to the Indonesian Parliament, Xi presided over a signing ceremony for the Morowali Industrial Park on the island of Sulawesi, along with Jokowi's predecessor Susilo Bambang Yudhoyono. The area was a small remote fishing village, without power and with few roads. The nearest major city was three hours' drive away. Yet it represented a key plank in Indonesia's strategy to derive more value from its natural resources. Over half of Indonesia's export revenues were dependent on natural resources – the country was the biggest exporter of thermal coal for power stations, as well as tin and palm oil. Chinese imports of raw nickel from Indonesia had soared from 161,000 tonnes in 2006 to 41 million tonnes in 2014. But politicians had long bristled at the sense of unfairness in the equation. China, with its rich access to credit, and the grand promise of Xi's BRI, provided a perfect opportunity to redress this balance. It was an example of how countries in South-East Asia were using China's external embrace to their own advantage.

A year later Yudhoyono launched Indonesia's first ban on nickel exports.

Tsingshan, with Xi's blessing, began to turn the Morowali Park into a giant stainless-steel factory, complete with a luxury four-star hotel, its own runway, a port, and a two-gigawatt-capacity coal-fired power plant. The Park employed some 38,000 workers. The project received hundreds of millions of dollars in funding from the state-owned policy lender China Development Bank (Beijing's key strategic lender) as well as loans from the China Export-Import Bank, and HSBC. The British bank trumpeted the park on its website, reposting a Chinese news article that said the lives of the villagers had 'been transformed: with power, there are more electric bulbs than fishermen's fires'.[21]

The pace of construction at Morowali was rapid, surprising every global steel producer, all of whom were only just emerging from the pains of the global financial crisis. 'Nobody ever imagined that a Chinese investment in Indonesia could have been so disruptive,' one Western stainless-steel producer told me. 'They have created the same capacity as the whole of Europe in a country where there is no consumption.' The company's stainless-steel production would destroy European industry, he said, even though it produced five times more carbon dioxide from its use of coal-fired power. 'They are destroying the environment,' he told me. Tsingshan went from having less than five percent of global stainless-steel production in 2009 to twenty-five percent, becoming so big that even Beijing threatened its Indonesian exports of stainless steel with tariffs in March 2019. Tsingshan had become the cheapest producer of stainless steel in the world.

Key to its success was the access to coal-fired power and cheap supplies of nickel. One of the company's partners in Indonesia told me they paid six cents per kilowatt hour for electricity, compared to ten to twelve cents per kWh in China. Because domestic miners could no longer export their nickel, Tsingshan was in a powerful position to dictate prices to local miners, since it was the biggest buyer in town. The industrial park consumed around 20 million tonnes of nickel a year. Tsingshan paid around $38 a tonne for the local nickel ore, compared to the roughly $65 a tonne that Chinese producers at home paid for material from the Philippines.[22] As the scholar Alvin Camba put it, the industrial park had created an oligopsony, where numerous miners competed to sell to a few buyers. As a result, the mining companies had less money to spend on environmental protection. Mining groups were 'cutting corners in order to make up for lost profit, passing socio-environmental costs onto Indonesian communities and environments', Camba wrote.[23]

The move into Indonesia also put Tsingshan in pole position to capitalise on the rise of electric cars, which by 2020 were using ever more nickel in their batteries instead of cobalt from the Congo. 'He's [Xiang's] always thinking five years out; he's not a day-to-day guy,' Lennon, the Macquarie analyst, told me. Tsingshan began converting some of its nickel pig iron used to make steel into a form suitable for batteries. Moreover, the Morowali Industrial Park paved the way for a host of Chinese companies to build processing plants to turn raw nickel into battery materials. Tsingshan entered a consortium with China's largest battery producer CATL and battery recycler GEM Co., as well as the Japanese trading company Hanwa. The high-pressure acid leaching plant that the consortium planned to build would process nickel and cobalt from raw ore. 'He [Xiang] will go at that with the same determination that he's built his stainless steel,' the person who knew Xiang told me.

Other Chinese projects followed the trail blazed by Tsingshan in Indonesia. On Obi Island, a joint venture between China's Ningbo Lygend and Indonesia's Harita Group had also built a processing plant to make nickel for batteries – and signed a deal to supply GEM with nickel. Huayou Cobalt also announced investments in two separate high-pressure acid leaching plants. By 2021 there were a total of eight Chinese projects under construction in Indonesia. The Chinese could build the plants so cheaply that they could make their money back within two years, whereas it would take a Western company up to fifteen years, according to nickel analyst Lennon.

As they watched Chinese companies rush to build projects in Indonesia, Tesla and the world's largest EV battery producers became worried. The reliance on coal-fired power meant nickel for batteries produced in Indonesia could be up to five times as carbon intensive as that mined in Australia or Canada, in terms of carbon dioxide emissions, according to Benchmark Mineral Intelligence. The sheer amount of energy needed to separate the nickel from the ore also meant that even if renewables were used, the number of solar panels required would mean a large land footprint – which could contribute to deforestation, which mining had already exacerbated.

Mining nickel in Indonesia required stripping large areas of the upland forest to access the ore near the surface. 'These nickel mines have very large footprints,' Steven Brown, who had previously worked in mining in Indonesia, told me. Erosion and heavy tropical rainfall led to run-off into the sea, impacting downstream communities. In addition, many nickel mines also produced a toxic pollutant, called hexavalent chromium, which could damage human health, he said. The chemical was the same one that Erin Brockovich had campaigned about in the US, a story that was made into a major Hollywood film. 'Since open-pit mining mostly takes place in the uplands, it generates downstream effects for rural populations whose farm land and residential areas are in the lowlands,' Arianto Sangadji, an Indonesian scholar based at York University, told me. 'In the rainy season, rivers flow across the landscape with brown alluvial sediment affecting small-scale agricultural lands. Floods become a normal event.' Floods damaged public facilities, paddy fields and gardens, he told me. Brown showed me two photos he had taken before and after nickel mining on one island, which showed a large plume of muddy outflow into the sea water.

Most of the nickel projects in Indonesia were also located in some of the most biologically rich and biodiverse areas in the world where logging and palm oil had already led to deforestation and conflicts with local communities.[24] The island of Sulawesi was home to seventeen unique species of primates such as the crested black macaque and the tarsier. It had been called a 'world-renowned laboratory of evolutionary biology'.[25] One study said that if the current rate of deforestation continued in Indonesia 'it could be catastrophic for the island's remaining wildlife and natural ecosystem services'. 'Mining is a persistent threat to primates and their habitats,' it said.[26]

Pius Ginting, a local activist, told me that the nickel plants at Morowali had already polluted local waters and threatened the lives of coastal communities that relied on fishing. One such group, the Bajo people, had been settled in a hamlet in Morowali in 1993 at the request of the local government. Their village had since been transformed into a major industrial area, where coal dust lingered in the air. Dumping of wastewater into the sea from the Morowali Industrial Park had made it no longer productive for fishing. In addition, coastal waters were also contaminated by red soil washed down by the rivers from mining sites. As a result, the fishermen needed to go out further into the sea to catch fish.

The greatest worry, however, was what all the nickel processing plants would do with their waste. Converting rock that contained about one percent nickel into a form suitable for batteries produced huge amounts of waste rock, which was laden with chemicals. If all the announced projects in the North Maluku and Central Sulawesi areas went into production, they would generate around fifty million tonnes of waste a year, according to Brown, a former employee of nickel miner Vale. Tsingshan and other Chinese companies had applied to the Indonesian government for permits to dump that waste into the sea, just as the Chinese mining company had done in Papua New Guinea. That could threaten local marine life, in an area known as the Coral Triangle for its rich coral reserves.

Indonesia had achieved its objective of attracting more manufacturing to its shores by leveraging its rich nickel resources. 'This is our country's business strategy, designed so we will become a major hub for the electric vehicle industry,' Jokowi said in 2020. It was a policy that harked back to the hopes of the developing countries at the Bandung Conference held in Indonesia in 1955, which had attempted to create a post-colonial world where resource-rich countries got more of the value of their resources. Indonesia, with its rich nickel reserves, had achieved more than most. 'With such large [nickel reserves], we can see that Indonesia has a strong bargaining position,' Jokowi's right-hand man Luhut said in June 2021.[27] Jokowi managed to get Korea's LG Chem to agree to build a $1.2 billion battery plant near Jakarta. Jokowi had also issued a personal invitation to Musk to invest in the country. The two spoke on the phone in December 2020, and Jokowi even went as far as offering Indonesia as a potential launch site for Musk's rocket company SpaceX.

But in its rush to attract investment Jokowi had ignored the environmental toll of Indonesia's nickel mining and processing. The country had lowered environmental standards in its determination to become part of the EV supply chain, the scholar Sangadji told me. In 2021 amid the spread of Covid-19 the Indonesian government made it easier to start projects even before an environmental impact assessment had been approved. Unless Indonesia forced its nickel industry to clean up, it would shoulder a growing environmental burden from the EV revolution. It would also help China to offshore its polluting industries, but desecrate its own land, water and air. 'The net result is we have clean air in our cities – but then we destroy a rich biodiversity area,' Pius Ginting said.

11

The Green Copper Tycoon

'Who benefitted the most from the viral disruptive outbreak of the [internal combustion] automobile? It wasn't the car manufacturers ... It was John D. Rockefeller and the guys in the oil business.'

Robert Friedland[1]

Our jeep drove down into the Kamoa-Kakula mine, and we moved suddenly from the bright African sunlight into the complete, enveloping darkness. It took a moment for my eyes to adjust. My driver, a white South African with neat short back and sides, barely skipped a beat as he drove on, moving us ever further downwards, on a winding path deeper into the bowels of the Congolese earth. Water dripped off the rock walls, where plastic ventilation tubes shuddered with air. It felt warm and muggy inside the mine; the air was thinner. I felt a momentary pang of claustrophobia thinking of the earth above me and the narrow entrance we had driven down. How was it possible, I wondered, for hundreds of people to work for hours down here, never seeing the brilliant sun overhead, never feeling the hot dusty day? A cement mixer passed us by and my driver honked his horn hard. 'I've seen a lot of accidents in my time,' he told me. He had once worked in South Africa's famously deep gold mines, doing ten years in the emergency rescue team. The depth of this mine was nothing compared to those ones, yet still I felt the enormous scale of humankind's quest to dig the earth. We soon arrived at the rock face where men worked on raised platforms drilling into the dark rock, the lights on their helmets breaking the darkness. The driver stopped our Toyota Land Cruiser and we put our helmets on, clipping on flashlights. I got out of the vehicle and tried to look at the rock – the ore body looked like a neat perfect square. My guide told me that workers were rushing to reach the richest part of the deposit in time for a visit by the president of China's CITIC Metal, a state-owned company that was part of one of China's oldest conglomerates. 'Every day, two metres, that's all I'm focused on,' he said. 'Meeting the target.' The president was likely to be very happy, I thought: this mine was one of the richest copper deposits on earth.

We had driven down bumpy dusty roads from the one-storey airport in Kolwezi escorted by an armed guard in a black beret to reach the camp in the afternoon light. The red dust swept up onto our windscreen like a storm cloud, obscuring the view as we dipped and dived to avoid ruts in the earth. We passed street vendors coated in dust, and had to veer to avoid potholes and large trucks carrying sulphuric acid from the nearby mines. This road was especially bad: we had been unable to take the quicker route through one of Glencore's mines as artisanal miners were throwing stones at people. Soon we were on a small country road, surrounded by long grasses, and the countryside seemed to open out before us. The mine camp appeared out of the dry bush just as darkness was arriving, an oasis of calm. 'Welcome', and '欢迎', a sign said outside in English and Chinese characters, next to the original drill bit that had found the deposit. A guard came to carry out a breathalyser test of the bus driver while we all waited. We then rolled into the campsite, a series of one-storey buildings and rows of huts that was ringed with barbed wire.

The camp was its own little world, separated from the darkness by bright lights and the bustle of men and women. Many of the geologists and other foreign staff had spent years living here (one couple had even raised their child in the camp, I was told). At

dinner that evening, I met some of them, as we sat around a bonfire
and ate meat from a barbecue that had been prepared. A makeshift
bar had been set up serving cold Congolese beer. The geologists
were all beaming as they drank – they were united by the feeling
they had helped find something unique in mining history – a rich
deposit of copper in the heart of Africa. It was the prize many had
only dreamt about earlier in their careers; most geologists never
discovered anything that turned into a mine. The mine would also
come into operation just as the world needed more copper: for all
the wind turbines, electric cars and charging stations required to
move beyond fossil fuels. It was like finding gold before the rush.

Someone pulled out a bottle of Chinese baijiu liquor to drink
and insisted we all do shots. I sat down next to Abraham, an outgoing Chinese man who worked for Zijin Mining, one of China's
largest gold miners. He had spent his whole career in Africa, and
spoke about mineral deposits with a sense of awe and passion.
Behind him sat his assistant, a tall thin man from Mali who spoke
fluent Chinese and told me he had studied in Shenyang on a schol-
arship. The richness of copper here was without parallel, Abraham
told me, sweeping his arms over the area. Chinese companies had
spent around $56 billion on overseas copper mines during the
2000s, but much of it had been wasted on low quality projects.
Zijin itself had spent $390 million buying a copper mine in Serbia
in 2019, but Abraham dismissed the deal as 'strategic' and just part
of Beijing's policy of winning over eastern Europe (Zijin would
later be forced to stop work on the mine after failing to comply
with environmental standards in the country). But this rich mine
in the Congo was different, he told me. It was potentially the largest discovery ever on the continent of Africa. And Zijin knew how
to operate copper mines cheaper than anyone else, he said proudly.

The richness of the deposit was something emphasised to me
the next morning over breakfast in the canteen, where workers in
their overalls ate omelette and drank instant coffee before their

shifts. I sat next to David Broughton, a geologist who had helped discover the Kamoa-Kakula deposit. Mining is a business that involves digging up rock that contains small amounts of valuable metals. Mining companies are essentially in the waste production business. So, any deposit with a higher grade of metal meant less waste rock, less energy and a smaller environmental footprint. The Kamoa mine boasted grades of six percent copper in the rock, compared to less than one percent in Chile, the world's largest producer, he told me. One drill area was called the 'bonanza zone', for its copper grades as high as eleven percent. Higher grades meant more copper for the same amount of effort. Given the huge amount of money miners spent on diesel-powered trucks as well as processing equipment, the importance of grade could not be overstated. 'Grade is king,' Broughton said.

A Canadian with a PhD from the Colorado School of Mines, Broughton had helped discover Zambia's largest copper mine before coming to the neighbouring Congo. He still lived in the camp and helped drill for more copper. The size of the deposit was forever expanding. In the afternoon we drove out to look at the drill rigs. It was a windy, sunny day and children played by the side of the red dirt road as we drove through villages of brick and thatched-roofed huts that used discarded mine materials as doors. We stopped at what the geologists said was the most exciting drill hole, Drill Hole 1450, which consisted of a blue pipe sticking out of the mud. Here the geologists had found copper at thirteen percent grade just 190 metres below the surface – an astonishing amount of copper that was easily accessible. The deposit was as thick as a ten-storey building. The drilling had even found copper grades of up to forty percent. 'This is one of the best drill holes anywhere on the planet,' Alex Pickard, a young Englishman who worked for Ivanhoe Mines, told me. 'We're standing on a few million tonnes of ore,' an English geologist, Tim Brooks, whom I had met the previous evening, chimed in. The geologists worked twelve-hour shifts manning the drill sites, removing tubes of chal-copyrite rock core from the earth and putting them in neat metal containers. The rock, which was flecked a golden colour with copper, was cut in half with a diamond saw and then into chips to be sent to a laboratory in Australia for examination. Drilling was an expensive business: the drills burnt through around 300 litres of diesel a day.

It had taken a long time to get to this point – almost thirty years. We take mines for granted, yet they are huge feats of engineering, discovery and money, that take time to bring to fruition. Developing a new copper mine takes at least ten years. To go from drilling holes to building an economic mine is a process requiring relentless optimism, hard work and hope. The man behind this discovery had those qualities in spades: Robert Friedland, a billionaire Canadian-American miner and film producer, whose credits included *Crazy Rich Asians*. A university friend of Apple's Steve Jobs, Friedland had made his career dis-covering, developing and then selling mines in far-flung corners of the world. He had helped discover one of the largest nickel deposits in Canada in the early 1990s, and then the giant Oyu Tolgoi copper mine in Mongolia a decade later. And here off a dusty road in the Congo he had found a third. 'He's a genius at dealing and a genius at finding or getting involved in very suc-cessful mining projects,' Norman Keevil, the former chairman of Canada's Teck Resources who has known Friedland for thirty years, told me. 'The odd person finds one discovery and makes a career out of that discovery but there are very few people who can do it more than once.'

*

Copper is everywhere in our lives, just like fossil fuels. It's in our dishwashers, our air conditioners, the pipes in our homes, the wires that bring us our electricity, and in our cars and phones. The length of electrical wires in our cars has grown to more than four kilometres, compared to a few hundred metres just thirty years ago, due to the increased use of electronic components. We are set to consume more copper over the next twenty-five years than we have over the past five thousand years as the global population increases and gets richer. We will need more cars, houses, refrigerators, air conditioners and buildings – all of which use copper (as well as lots of other materials such as steel and plastic, of course). Much of the growth will happen in cities, with the growth in building stock set to double by 2060.[2] That's the equivalent of putting up another New York City every month for the next forty years. Copper's role in our lives is so ubiquitous that the metal is known as Dr Copper, for its ability to predict the health of the global economy. If you need a quick check of where the global economy is headed, then the copper price is a good guide.

The clean energy revolution will only increase this thirst for copper. Ever since Thomas Edison built the first power generation station in 1882, we have used copper for electrification due to its high conductivity and malleability (only silver is a better conductor, but it's much more expensive). Edison displayed a cubic foot of copper in his study, which was a gift from the copper industry to show their appreciation. As we move to using electricity rather than fossil fuels, and renewable sources such as wind and solar rather than coal, we will need ever more of the metal for wiring. Global electricity supply is set to double or even triple by 2050, as we shift to electric cars and to using electricity to make things like steel.[3] An electric vehicle can use up to three-and-a-half times more copper than a petrol car: there are between forty and eighty kilograms of copper in an EV. (That of course increases with the size of the electric vehicle – an electric bus can use between eleven and sixteen times more copper than a petrol bus.) The metal is found everywhere – as copper foil in the battery, the electric motor, the inverter and the wiring. There was perhaps a mile of copper in my Tesla. And that was just my car – 30,000 electric vehicles would consume the same amount of copper as a skyscraper. If we wanted to produce 300 million electric vehicles – a third of the current global fleet – we would need 20 million tonnes of copper, almost the equivalent of total world consumption.[4]

Producing enough copper for a global shift to clean energy will require new sources of supply. But where will it come from and who will mine it? Chile's copper mines have been in business for over a hundred years, requiring miners to dig ever deeper to find copper ore. Since 2005 Chilean copper mines have had to double investment just to sustain production at that level due to the decline in ore grade. The grade of copper in Chile had fallen from 1% in 2004 to around 0.67% in 2019. As a result mining required more energy and generated more waste. Between 2001 and 2017 in Chile, fuel consumption increased by 130% and energy consumption by 32% per unit of mined copper, due to the decreasing ore grade.[5] Friedland was dismissive of Chile's copper mines, calling them 'little old ladies lying in bed waiting to die'.[6]

Every day the price of copper is set on the LME, which was founded in 1877. In the nineteenth century metals merchants would draw a circle in the sawdust of a London coffee house around which they would trade metals. Later the exchange would prove pivotal in hedging copper shipments for the three months it took to ship the metal from Chile around Cape Horn on the tall-masted sailing clippers. The copper was often shipped to Wales, where it was refined in the Lower Swansea Valley – a process that sullied the area for decades with contamination. Today traders, almost all young men in tight-fitting suits, shout orders and use hand gestures around a red sofa known as the 'Ring' in the centre of London. The trading sessions are short and frenetic and reminiscent of an older era of open-outcry trading, which has mostly shut down in Europe and America. In 2012 the Hong Kong Stock

Exchange bought the LME for £1.4 billion, giving China greater control over how metals are traded. That made sense: on almost any given day the price of copper is determined by what happens in China. When China is on holiday, trading volumes in copper on the LME drop significantly. China's hedge funds are fast movers in the copper market, trading the arbitrage between prices on the Shanghai Futures Exchange against those in London. By the time a trader wakes up in London and has breakfast they have often missed the action in the copper market. Being a successful copper trader today means having a unique insight into Chinese policy.

China completely dominates the global copper market, and is responsible for around half of global consumption. No country even comes close. Copper that is dug up in Africa or South America is shipped to China to be refined, then fabricated into products that end up in buildings, air conditioners or electronic products for export. Yet China produces only eight percent of the global supply of copper, meaning it is almost wholly reliant on foreign supply. As a result, Chinese copper miners are constantly on the look-out for overseas acquisitions. Beijing – as in other commodities – has shown little confidence in relying solely on a global free market to provide supply.

But as Chinese companies looked outside their borders in the early 2000s for copper to feed their growing economy, they found that the big copper mines had already been bought up by the large Western mining companies. In Chile BHP, the Anglo-Australian miner, owned the world's largest copper mine, Escondida, which was 3,000 metres above sea level in the Atacama Desert. American copper miner Freeport-McMoRan owned the giant Grasberg copper and gold mine in Indonesia, as well as the Cerro Verde mine in Peru. Finding a new copper mine was not easy so China had to buy what it could. In 2014 a consortium led by state-owned miner Minmetals bought the Las Bambas copper mine in Peru from Glencore for $5.85 billion, in part because Glencore needed to sell it to get Chinese government approval for its takeover of rival Xstrata.

The Congo offered China an answer. Not only did it have rich resources of cobalt for batteries, but it was also Africa's biggest copper producer. Western mining companies, with the exception of Glencore and Freeport, saw the country as too risky and too corrupt to do business in. But as it became clear demand for copper would only increase with the rise of electric vehicles and renewable energy, China eyed an opportunity. Following Kabila's victory in the country's first democratic elections in over four decades in 2006, China stepped in to provide a large package of funding to the country. Kabila wanted a 'Marshall Plan' effort to rebuild the broken country, and Chinese money didn't come with any lectures. The $9 billion deal with the state-owned China Railway Engineering Corporation provided mining licences in Kolwezi in return for spending on infrastructure. It was so large its value exceeded the government's budget during the year it was signed. But beyond the headline figure the project was dogged by years of difficulties and ultimately failed to give China a huge source of copper. The next move by China was better executed and came during a fall in copper prices that started in 2014. That price plunge put Western mining companies, and their shareholders, under growing stress.

The Tenke Fungurume deposit was spoken of in almost mystical terms by the geologist John Guilbert to his graduate students at Arizona University in the 1990s. It was a copper and cobalt mine lying undisturbed in Mobutu Sese Seko's Zaire. When Adolf Lundin, a Swedish oil and gas entrepreneur who had made his money in Qatar, asked Guilbert in 1994 what was the best copper mine on the planet he replied: 'Tenke.' Two years later Lundin, whose family motto was 'No guts no glory', reached out to Mobutu to begin discussions over Tenke, acquiring the site for $250 million in agreement with the state-owned mining company

Gécamines. A year later after Mobutu was overthrown Lundin gave the rebel leader Kabila $50 million to try and secure the mine as the country was plunged into civil war. It would take until 2005 for work to begin on the mine, this time with US copper miner Phelps Dodge, which was taken over by Arizona-based miner Freeport-McMoRan the following year in a $26 billion deal that created the world's largest copper miner.

Richard Adkerson, the chief executive of Freeport-McMoRan, had been born in rural Mississippi to a family of small farmers. He was enamoured by the giant Tenke Fungurume mine and decided to spend as much as possible to develop the deposit. A private runway was built and a highway to export copper and cobalt.[7] But in 2015, commodity prices had plunged due to a weakening of demand in China and Adkerson was facing pressure from the billionaire New York-based activist investor Carl Icahn, who had become the largest shareholder of Freeport a year earlier. Icahn wanted the miner, which traced its roots to 1912, to reduce its debt levels, which had jumped to $20 billion after a series of ill-timed oil and gas acquisitions. Freeport's shares were in free fall. During the LME week in late 2015 I watched as Adkerson took the stage, grabbed a microphone and sang 'If you're going through hell, Keep on going,' a song by Rodney Adkins, at the Intercontinental Hotel in Park Lane. Gone were the oysters and the large arrays of sushi stations. Instead, delegates picked at marshmallows dipped in chocolate.

At the same time, China Molybdenum, a Hong Kong and Shanghai-listed miner that had its roots as a state-owned enterprise in the central Chinese city of Luoyang that was founded in 1969, was looking to expand internationally. Yu Bo, a billionaire former trader who had invested his fortune in China Moly when it restructured into a private company, had high ambitions. He had increasingly taken on the trappings of the global elite: he had a private Airbus 319 for his wife and mistress, a Rolls-Royce and a penthouse in Vancouver, according to former employees. He had already bought the Northparkes copper mine in Australia and a niobium and phosphate mine from Anglo American in Brazil over the past few years. The company stressed to the media that it was private and had nothing to do with the Chinese state. Its international arm was called China Moly International and was headed by a soft-spoken Indian-American Phelps Dodge veteran called Kalidas Madhavpeddi.

In early 2016 China Moly began talks with Freeport over Tenke, with the help of Citigroup. Yu also teamed up with the Chinese state-backed private equity company, BHR Partners, where President Joe Biden's son Hunter Biden was a director, on the deal. By November of 2016 China Moly agreed to pay $2.65 billion in cash for Freeport's stake. Shares in Freeport rose on the news and Icahn sold out some of his stake a year later. Freeport believed that China Moly would retain a lot of its workers and had pledged to help with the transition. Yet Yu quickly asserted his control. Despite his talk of building an international mining company, he quickly became paranoid of foreigners, according to former workers. A suggestion was made to install video cameras in every overseas office to monitor foreign workers. And a new human resources employee who many suspected of being a spy travelled to each overseas office requiring lists of everyone working there and details of their ages and salaries. Madhavpeddi soon stepped down. China Moly also cemented its control, buying out BHR. The US had ceded one of the world's largest copper and cobalt deposits – to satisfy a billionaire hedge fund. By 2021 there were no US mining companies with deposits in the Congo. 'There is no equal masterplan in the Western world, the US has ceded Africa by pretty much not being there,' one former China Moly employee said.

China Moly became a key part of China's electric vehicle and battery supply chain. In mid-December 2020 it acquired a further copper and cobalt deposit nearby owned by Freeport, Kisanfu, for $550 million. The deposit contained an estimated 3.1 million tonnes of cobalt and 6.2 million tonnes of copper. The following year a subsidiary of China's largest battery maker, CATL, bought a stake of over twenty percent in the project for $138 million. The deal happened just as President Joe Biden ordered a survey of the US supply chains, promising to reduce dependence on China.

Following the Tenke purchase the Chinese also found a perfect opportunity in Friedland.

*

I first met Friedland at a copper conference in Santiago in early 2016 when he put up a photo of Tesla's new mass-market Model 3 electric car and heralded it as the future for the mining industry. 'Copper is the king of metals,' Friedland told attendees at the conference in Chile, the world's largest copper producer. 'Just based on world ecological and environmental problems every single solution drives you to copper – solar power, wind power, electric cars, you name it.' His speech was the first sign that something new was stirring in the staid mining industry – a transition to electric cars and renewable energy that was due to boost demand for metals significantly.

Friedland's speeches were carefully crafted productions that were incredibly successful at getting investors to part with their cash. Normally miners drily go through their slide deck, with details of their projects, how low-cost they are and how they're going to come to the market at just the right time in the commodity cycle. But Friedland goes for the jugular – his mines are going to save humanity from our predicament. He normally started with a video of the American singer and satirist Tom Lehrer singing about pollution in 1967. Then he told you how metals would be at the core of any solution. At some point, without you much noticing, he switched into salesman mode and started to mention all his projects, which just happened to have the right metals.

'Opening in a theatre near you is the revenge of the miners,' he said on a cold day in north London in November 2017 as I sat in the audience of the bland Business Design Centre in Islington. The copper price would go so high you would need a telescope to see it, he added. Whenever I met him, I always found Friedland mesmerising, though I knew I was just succumbing to his well-practised intensity and charm. But in a mining industry run by white Australians and South Africans who kept to the script, he was refreshing.

Born in Chicago to parents who emigrated from Germany, Friedland studied at Reed College in Oregon. Naturally charismatic, he was popular immediately and ran for student president in his first year. It was at Reed that Friedland met a young Steve Jobs. The two became close friends and Friedland was 'one of the few people in Jobs's life who were able to mesmerize him', his biographer wrote.[8] Jobs thought of Friedland, who was four years older, as a sort of guru and adopted some of his key charismatic traits, including the famous 'reality distortion field' that he used to compel others to do what they thought was impossible. They were both into eastern religions and spirituality. Friedland had travelled to India in the summer of 1973 to meet a famous Hindu guru Neem Karoli Baba and Jobs took the same trip a year later. Friedland was a father figure to Jobs and a seeker 'in the tradition of Ram Dass', the American spiritual teacher, according to Jobs's former girlfriend Chrisann Brennan. 'He had started out exploring the intellectual traditions of consciousness, then moved to LSD, and from there became a student of Neem Karoli Baba and the mysticism found in the East,' she wrote.[9]

After he graduated in 1974 Friedland moved full-time to an apple farm owned by his uncle in Switzerland, which was around forty miles south-west of Portland near McMinnville. The farm became a hippy commune where groups of young people from the Hare Krishna temple worked the apple orchard, meditated, and ate vegetarian food together. Friedland took the name of Sita Ram Dass and looked 'like a Caucasian Krishna'.[10] Jobs, who had dropped out of Reed, worked at the orchard, helping to produce cider.

Pretty soon, however, Friedland started operating the commune as a business, according to Jobs. 'It started to get very materialistic,' Jobs told his biographer. 'Everybody got the idea they were working very hard for Robert's farm, and one by one they started to leave. I got pretty sick of it.'[11]

Farming also soon lost its interest for Friedland and by the late 1970s – and with the backing of some Vancouver brokers – he moved on to gold mining, promoting a series of ventures on the Canadian stock markets, many of which failed to take off. Friedland discovered a natural and unrivalled talent for promoting mining ventures. In the 1980s 'Friedland moved the art of the penny-stock promotion from vaudeville to Broadway,' as one author wrote.[12] One of his ventures earned him the nickname 'toxic Bob', after heavy metals and effluent from the Summitville gold mine in Colorado leaked into a nearby river in 1992, leaving the government with a large clean-up bill. Friedland has said the Summitville mine was not the sole cause of the water-quality problem and that the sources of the heavy-metal contaminants were due to past and other mining in the area as well as natural acidic runoff. In 2001 Friedland paid $20.2 million as part of a legal settlement with the US government. (The US government also made a payment of $1.25 million to Friedland as compensation for costs resulting from its efforts in 1996 to seize his financial assets in Canada.)

Friedland's first big success was the discovery of the Fort Knox gold deposit in Alaska in 1987, which was sold to Amax Gold in 1991. A few years later his geologists stumbled upon a huge nickel deposit in Voisey's Bay in Canada while looking for base metals and diamonds. Friedland played two corporate suitors off against each other to sell the mine for $3.2 billion to the nickel giant Inco in 1996. It was a sales masterstroke and made Friedland a fortune overnight.

From his base in Singapore Friedland then rapidly expanded across the globe, targeting Asia. 'Our big idea is, if the consumption is in Asia and the demand is in Asia, why not engage in mineral exploration and development where the market is?' he said in 1996.[13] He entered a copper mining project with the military government in Burma and bought exploration rights in Mongolia from BHP for just $5 million. The copper and gold mine his company subsequently discovered on the final drill they did in Mongolia, Oyu Tolgoi, would go on to become one of the biggest mines in the world and was bought by the mining giant Rio Tinto. 'He's willing to go where others have given up on,' Norman MacDonald, a fund manager at Invesco in Toronto who grew up in the nickel town of Sudbury, says. 'Mongolia was the same situation. There's very little competition … across the board that entrepreneurial spirit and talent is dwindling in the mining business.'

He first went to the Congo in 1996, just as a rebel movement based in the east of the country had pledged to overturn the three-decade-long rule of Mobutu. He visited the Congo's capital of Kinshasa, where he stayed at the Intercontinental Hotel with his second wife, Darlene. There were mortar bombs coming across the Congo River from neighbouring Congo-Brazzaville. 'There was no electricity. There were lots of dead bodies,' Friedland recalled. He had been introduced to the rebel leader Laurent Kabila, who was advancing with the help of military forces from Rwanda and Uganda, by Isabel Dos Santos, an Angolan businesswoman who was the daughter of the country's president at the time. Friedland went on TV to defend Kabila's advance to power, and in return got 14,000 square kilometres of land outside the copper mining town of Kolwezi. 'We knew Congo was the world's richest source of copper,' he told me.

Stretched across the south-east of the DRC into Zambia, the copper belt is one of the world's richest sources of both that metal and cobalt. As we saw in earlier chapters, the ground around the town of Kolwezi is so rich in minerals that copper and cobalt can be dug up from near the surface, without the need for a deep mechanical mine. The area was mined for hundreds of years before the Belgians arrived, with copper exported by African and Portuguese traders to the Atlantic coast. As the historian Miles Larmer wrote, the copper belt provided the basis for the rise of major societies such as the Luba and Lunda kingdoms in the seventeenth and eighteenth centuries. 'Mineral exports enabled the import of technology and knowledge, which raised population density, expanded land under harvest, and strengthened central state capacity to enslave subject peoples and extract tribute from areas such as the zone to the south and east of the core Lunda kingdom where copper was mined,' he wrote.[14] Mined by the Belgians at the beginning of the twentieth century, copper from the city of Kolwezi was used in shells fired in battle during the First World War in France. The Congo became one of the largest producers in the world in the 1960s, with output peaking in 1976. But by 1995, production was down by ninety percent. The industry had fallen further into disrepair under Mobutu.

Friedland's land lay to the other side of Kolwezi from where the Belgians had mined. It had not been explored because it lacked any surface indications of copper, such as a break in the vegetation, or the distinctive bright turquoise malachite rock that contained copper. These indications were known as 'showings' by geologists. Malachite rock has a low melting point so was melted for thousands of years to produce copper in the Congo and Zambia, with the metal often used as a form of currency. 'When the Europeans came here a lot of the deposits were already known,' Broughton told me. 'There wasn't much discovery; it was more about showing the locals we're interested in malachite and asking where do we go?' The earliest maps from the 1900s, produced by the Europeans, already had the large mines in the area on them. Yet the Copperbelt was thought to end around the town of Kolwezi. But Friedland had a hunch that his area might contain copper.

Following the end of Congo's civil war in 2002 geologists started exploring the area, sampling sediment in local streams, and mapping the area using an airborne magnetic survey. In 2008 Broughton joined Ivanhoe and started to focus on the Kamoa area, which was one of many targets. That year, geologists began to drill the deposit. They lived in tents on the site and braved the snakes and rainy season. 'The financial crisis had melted everything else down ... we were the only company that stayed [in the DRC]. Drilling companies phoned me up saying: "Do you need any drills?",' recalled Broughton. They continued to drill into the rainy season of 2009 to February, announcing the discovery of a large orebody in April of that year. 'We were living in tents and mud ... But when you make a discovery like that you're in the clouds for two years.'

Friedland announced the discovery in April 2009 in the capital of the world's biggest copper producer – Santiago, Chile. 'It's thousands to one,' Broughton said. 'It's a similar success ratio as the pharmaceutical industry. Only a few geologists in their career find something that turns into a mine.' All Friedland needed now was $1.1 billion to build the mine.

Friedland first went to China in 1981 and has maintained close links to the country ever since. In the 1990s he entered a partnership to build six-storey plastic apartment buildings in Shanghai with a highly connected Communist Party organisation

– the China Disabled Persons' Federation. The federation was at the time headed by Deng Pufang, the son of former leader Deng Xiaoping, who became disabled after being pushed out of a building during Mao Zedong's chaotic and bloody Cultural Revolution in the 1960s.[15] In the same decade Friedland explored for gold in south-east Fujian province and met Chen Jinghe, who would go on to become head of Zijin Mining, China's largest gold producer, which is listed in Hong Kong with a market capitalisation of HK$80 billion.

As copper prices plummeted in 2015, Zijin agreed to invest $412 million to acquire a fifty percent interest in the Kamoa mine. Zijin also bought shares in Friedland's company Ivanhoe Mines – which was listed in Toronto. Three years later CITIC Metal, a unit of the Chinese state-owned conglomerate that has $900 billion in assets, bought a twenty percent stake in Ivanhoe for $556 million – increasing it to twenty-nine percent a year later.[16] 'CITIC Metal's strategic investment in Ivanhoe reflects our strong belief that the long-term prospects for the mining industry are extremely positive, especially given the projected robust future demand for key metals, such as copper, for renewable energy infrastructure and electric vehicles,' Miles Sun, president of CITIC, said at the signing event in Beijing.[17]

Friedland had provided China with one of the world's richest deposits of copper. The timing was perfect: as the Kamoa mine produced its first supplies of the metal in the summer of 2021, copper prices traded close to their highest level in ten years. 'There is such a thing as luck and luck is good,' Friedland, who was about to turn seventy-one, said from his home in Singapore. Paul Gait, one of the smartest metals analysts in London, called the discovery the 'most important geological development of our age'. Friedland would eventually cede full control to the Chinese, helping them to create a multinational mining company, he believed. 'What's driving him on is his legacy,' he said. 'It's no longer flipping assets but about building something for the future. He wants to be a partner with the Chinese and that process is going to make him exceedingly wealthy. You've got the synergy of his ability to operate, own and find mining assets, coupled with the balance sheet of China.'

12

The Final Frontier: Mining the Deep Sea

'The rush to mine this pristine and unexplored environment risks creating terrible impacts that cannot be reversed. We need to be guided by science when faced with decisions of such great environmental consequence.'

Sir David Attenborough[1]

'There are minerals in the deep sea that might be valuable to us, so millions are being spent on mining there. It's the biggest land grab in history, and most people don't even know it's happening.'

Ocean explorer Sylvia Earle[2]

In 2018 in a restaurant in the heart of the City of London, Gerard Barron brandished a small black rock – the size of the palm of his hand – and heralded it as the future. He placed the potato-sized rock on the clean white tablecloth, pausing to let me look at it, as if he had brought back a trophy from a far-flung tribe or a rock from the moon. I sensed he enjoyed this well-rehearsed set piece of theatre, especially in a smart restaurant, where we were a world away from mining. The Australian entrepreneur believed these rocks, formed over millions of years at the bottom of the ocean, could help satisfy the growing demand for the metals used in batteries and clean energy technologies, and were therefore critical to the transition away from fossil fuels. 'It's all here,' he told me. 'All the metals we need.'

Barron donned the familiar language of an environmentalist in decrying the destruction mining wreaks on our habitat and societies: deforestation, pollution and child labour, he listed them off for me. 'There's no point thinking you're doing the planet a favour by driving an electric vehicle if those materials have been mined by the hands of children,' he said, 'or you have had to deforest important rainforest assets.' He is keen to promote himself instead as a clean energy entrepreneur and has called his start-up 'DeepGreen'. The company's corporate brochure is replete with the familiar images of solar panels and wind turbines standing on empty landscapes, devoid of humans or any indus-try – an effort to distance himself completely from the dirt and reality of land-based mining.

The rock was a nodule from the deep sea, a place few humans have ever visited. It also contained – by geological luck played out over millions of years – just the metals needed for lithium-ion bat-teries. It's as if they were left there for just the point when we have fucked up the planet, Barron said. Less than twenty centimetres wide, the so-called nodules contained nickel, manganese, copper and cobalt – a lithium-ion battery lying on the ocean floor. Up to 38 million square kilometres of the ocean floor was covered with these nodules – an area more than twice the size of Russia.[3] DeepGreen, which was also backed by shipping group Maersk, had the rights to explore 74,713 square kilometres of the nodule-rich Clarion-Clipperton Zone in the Pacific, through agreements with the Pacific Island nations of Tonga, Kiribati and the small island republic of Nauru (which has a population of around 12,000). The zone, which lies between Hawaii and Mexico and is as wide as the United States, had enough nodules to electrify the global fleet 'several times over', according to DeepGreen (which later changed its name to The Metals Company).[4] Barron planned to use underwater machines to suck up thousands of tonnes of the nodules from the deep sea and send them up a pipe to a waiting ship to be taken to land to be processed. It was mining by deep-water hoovering.

Barron grew up on a dairy farm in Toowoomba, Australia, as one of five children. He started his first company while at the University of Southern Queensland where he studied economics and marketing. 'I already had four jobs, and I had no room for five jobs,' he recalled. He decided to make more money by heading out with his clipboard and becoming a financial salesman. He set up a company that helped connect companies to borrowers so they could refinance their debts during a period of high interest rates. By the time he graduated in 1986 he was driving a BMW with its own phone. He was hooked on becoming an entrepreneur and after graduation formed a number of companies. He came across a Chinese lead-acid battery company that was looking for a distributor in Europe and jumped at the chance, moving to Oxford when he was twenty-four. Barron settled into English life and played for the Blenheim Park Cricket Club. He also visited China, seeing first-hand the country's rapid industrialisation. Pudong business district in Shanghai was a 'sea of cranes' back then, he told me, while Changsha, now a big city, was a rural backwater. The Chinese battery company was edging out its competitors in Korea and Japan by making lead-acid car starter batteries that were cheaper. It was dirty work that was labour-intensive, far from the automated battery factories in today's China. Barron later returned to Australia and founded the digital advertising company Adstream in 2000, in which he sold a stake to Australian telecommunications giant Telstra six years later, making $20 million. The same year he started to invest in what was the first commercial deep sea mining venture, Nautilus Minerals, which was founded by mining journalist Julian Malnic in 1997. Malnic had taken the idea from an Australian scientist he had interviewed who had discovered minerals offshore of Papua New Guinea, in so-called hydrothermal vents – volcanic eruptions from the ocean floor. Malnic decided that rather than be a journalist he'd go and mine the vents himself and flew to PNG to buy the rights. Five years later the company was taken over by David Heydon, an Australian geologist who had started a flight-tracking software company, which went bust after the September 11 attacks. Heydon sold off his tennis court and swimming pool to fund the company, which listed on the Canadian stock exchange in 2006. Barron, who was friends with Heydon from Brisbane, invested in 2001, eventually turning a $226,000 investment into a reported $31 million, following a surge in its share price. 'I knew sufficiently little about metals that I thought this was an obvious and easy idea,' Barron said. When I asked him about the $31 million, Barron declined to confirm or deny it, insisting only that he 'did well'. He got out just in time: Nautilus filed for bankruptcy in late 2019, leaving public shareholders with nothing and PNG with debts representing around one-third of its annual health budget.*[5]

Barron wore an open, loose-fitting leather jacket and a V-necked white vest, and had long unruly hair that was sprinkled with grey. He also sported a scruffy goatee. On his wrist he had a colourful beaded thick wristband that gave him the air of a member of the Extinction Rebellion climate change protest movement. I inadvertently used the word miner to describe him, and he winced, saying he prefers 'harvester', in reference to the fact that he just aimed to suck up the millions of nodules that lie on the bottom of the ocean.

The US Geological Survey said the deep sea, which covers around half of the earth's surface, contains more nickel, cobalt and possibly rare earth metals than all land-based reserves combined.

---

* Barron says that Nautilus never wanted PNG to take equity in the company but the country insisted.

Access to the metals is governed by an international agreement called the Law of the Sea, and implemented by the International Seabed Authority, which is in charge of regulating any mining activities. In the Clarion-Clipperton Zone, where Barron held exploration licences, there were an estimated 21 billion tonnes of nodules.[6] The nodules also contained all the metals together, unlike land-based mines, which would lower the cost of extraction. 'We need the metals. Or we will have to stop doing practically everything we do – we have to stop the green technology, we have to stop having cell phones, electric cars,' James Hein, a senior scientist at the US Geological Survey, told me. 'We have an opportunity right now to do something. Let's be sure that when it is done, and it will be, that it's done in an absolutely environmentally sound way.'

The problem was the deep sea had never been mined and scientists were only beginning to understand how rich with life it was.

*

The ocean is one of the last unexplored wildernesses and one of the only areas of the planet apart from Antarctica not exploited for the extraction of raw materials. Unlike space, the deep sea holds little romance or mythological significance, operating rather as a lacuna in our knowledge. We look up to the stars at night to consider the vastness of the galaxies – but rarely consider what lies beneath us. We know more about Mars or Venus, even though the deep sea covers over half of the planet and may contain ninety percent of all life on earth. It's estimated that only 0.0001% of the deep sea has been sampled by scientists.[7] Few people have descended to the deepest oceans, apart from a couple of billionaires such as former private equity executive Victor Vescovo and the film director James Cameron. Cameron's footage showed a world shrouded in dark blue darkness that appears flat and lifeless. But far from being flat, the deep sea is covered with trenches – the Mariana Trench, the Puerto Rico Trench, the South Sandwich Trench, the Java Trench – as well as underwater mountains, canyons, crusts and volcanic hydrothermal vents that spew mineral-rich fluids. There are more mountains in the deep sea than on earth, according to oceanographer Greg Stone.

The deep sea lies in constant darkness, far below the twilight zone where showers of organic material, known as 'marine snow', fall from the shallows and sustain the food chain. It was long thought to be devoid of life. But the abundance and variety of life in the deep was confirmed in February 1977, when an expedition using the submersible *Alvin* north-east of the Galapagos Islands stumbled upon fantastic creatures living on deep sea hydrothermal vents. 'Isn't the deep ocean supposed to be like a desert?' the ship's geologist Jack Corliss asked from inside the *Alvin* using a telephone link to the ship on the surface. 'Well, there's all these animals down here.'[8] These included giant white clams, white crabs, giant tube worms and 'dandelions', orange-yellow balls that are suspended off the ocean floor. Today there are around 25,000 known deep-sea species, though some estimates are much higher since the figure represents only the tip of our understanding. 'Every sample we take tends to come up with some new species in it,' Phil Weaver, the founder of Seascape, a UK-based environmental consultancy, told me. Without sunlight life is sustained by organic debris from the upper oceans as well as chemosynthesis, where energy from chemical reactions is used to convert dissolved carbon dioxide into organic molecules – helping to sequester CO_2 absorbed by the oceans and preventing it being released into the atmosphere. Species include the 'gummy squirrel' and an albino octopus nick-named Casper, as well as multiple worms and snails. In just one type of worm, the polychaete worm, scientists found at least 300 new species.[9] A host of microbial life also lives on the polymetallic nodules that lie in sediment on the seabed, which have formed at a rate of millimetres per million years. It is thought that these microbes may have processed the metals from the seawater onto the nodules. Incirrate octopods, a new species discovered in the deep sea, have also been found looking after their eggs on dead sponges attached to nodules at depths of over 4,000 metres.[10]

Mining the seabed is focused on three distinct underwater terrains that each have different resources and characteristics: the fields of polymetallic nodules that lie up to 6,000 metres down on the sea floor; so-called massive sulphides, which are deposits that form along ridges or near hydrothermal vents – outlets where hot water escapes from hot volcanic rocks; and cobalt-rich crusts, which lie on the flanks of underwater mountains called seamounts. Of the three, the nodules host comparatively less marine life but have the greatest diversity, while vents have the highest abundance of life.

Mining for deep sea nodules will still require a large surface area, at almost 80 square kilometres per million tonnes of ore mined, compared with 0.52 square kilometres on land, according to Seascape consultants. It will churn up plumes of sediment and cause sound and light pollution that could impact local ecosystems.

Evidence showed that areas in the deep sea that were first explored in the 1970s and 1980s have still not recovered. Disturbance from experimental mining of nodules conducted in 1978 in the French area of the Clarion-Clipperton Zone at a depth of 5,000 meters was still visible decades later, for example, with a lower density of tiny worms called nematodes at the site. In a study published in 2020, researchers returned to the site of a 1989 study in the Peru Basin of the South Pacific, which had attempted to mimic deep sea mining by ploughing the 4,150-metre-deep sea floor repeatedly with a plough harrow over eleven square kilometres.[11] They found the plough tracks were still visible and that they were 'clearly devoid of nodules', which had been ploughed under or pushed to the sides of the tracks. Local microbial activity had been reduced by up to fourfold in the affected areas, and the researchers suggested it would take fifty years to return to pre-disturbance levels. These experiments represent only a fraction of the disturbance likely to be caused by a twenty-year mining operation, operating as much as possible in order to make a profit from volatile metal prices.

The UN Law of the Sea had protected the deep sea from mining activities for over twenty years, and no company has yet been allowed to start mining for minerals. NGOs such as Greenpeace and Conservation International believed that more time is needed for further study before mining can start. This call was backed by prominent naturalists such as David Attenborough. 'Do you gamble with those uncertainties in favour of the mining operations or do you give the benefit of any doubt to the protection of marine ecosystems?' David Santillo, a Greenpeace scientist at the University of Exeter, said to me. 'To me the latter is more important.' Louisa Casson, an ocean campaigner at Greenpeace, said deep sea mining risked the wholesale 'extinction of species'. Recycling of metals was a much better option, she said.

Barron acknowledged that deep sea mining would damage the seabed in the areas set aside for it. But he sought to portray it as better than mining on land, pointing to all the ills caused by thousands of years of mining, from deforestation to dumping of mine waste. He was not short of evidence on that front. He also correctly pointed out that almost all studies showed that recycling of metals would not be enough to meet demand. Some scientists also agreed with him that the only way to know the impact was to start mining. 'We are finding out a lot more but we can't answer the key questions about how these organisms will respond to the disturbance from deep sea mining without doing experimentation on the sea floor,' Dan Jones, a scientist at the National Oceanography Centre, told me.

Still, Barron's consistent promotion of seabed mining over land-based mining could at times feel a little forced. In one webinar in May 2020, consultants for DeepGreen flashed up photos of a cute-looking tarsier monkey from the Philippines, which they said was threatened by nickel mining in the country. Wouldn't we rather protect this human-like monkey than some worm in the deep sea we couldn't even picture in our heads? 'We have to get the metals from somewhere,' the consultant said. 'If we say "leave the deep sea alone" then we're participating in a great deal of damage on land, and extinctions of species that personally would matter more to people than a nematode that hasn't even been named or a worm or possibly a sea cucumber. I would prefer if I had to choose saving snow leopards, orangutans or the Philippine Tarsier ... that has more spiritual and aesthetic meaning to me than an invertebrate on the bottom of the ocean.'

In April 2020 the company produced a peer-reviewed study in a scientific journal that claimed the carbon footprint of deep-sea mining was lower than that of land-based mining.[12]

But was the comparison even fair? As James Hein of the US Geological Survey pointed out, 'it is difficult to compare the value of a terrestrial rainforest or grassland with that of a deep-sea ecosystem. What measures should be used? Ecosystem service, biodiversity of the system as a whole, operational CO_2 emissions or respective economic, social and environmental impacts?'[13] Because deep sea nodules contained many metals together, the CO_2 impact was likely to be lower than land mining. But land-based mining would continue, even if we mined the deep sea. It would just prevent some of the higher-cost mines on land from opening.

The hunt for clean energy minerals was only the latest iteration of a decades-long quest to mine the deep sea that had involved the CIA, Lockheed Martin and Barron's bankrupt Nautilus Minerals.

*

The potential bounty at the bottom of the sea was first realised as early as 1873 when HMS *Challenger*, a British ship, hauled up 'several peculiar black oval bodies which were composed of almost pure manganese oxide' while on a voyage to explore the Canary Islands in the North Atlantic. At the time manganese was used to bleach cotton for use in textiles, one of England's largest industries. 'Though the bottom of the sea at present could never be made a paying source of supply,' the occurrence of manganese there 'may turn out to be an important fact in geology,' the ship's chemist wrote prophetically.[14] Yet because demand was easily met by land-based resources it took almost 100 years before the minerals started to receive serious attention. Before the Second World War, the nodules were regarded as museum curiosities, the way moon rocks are now, according to the scholar Ole Sparenberg. It was not until 1965, when the American mining engineer John Mero published a book called *The Mineral Resources of the Sea*, that mining of the nodules was considered. Mero, like Barron decades later, argued that mining from the deep sea would avoid relying on unstable countries. While the earth's crust contained enough minerals to support any population 'the problem with ore deposits … is not the total quantity available in the continental rocks but their uneven distribution and mankind's propensity for indulging in political and economic systems which inhibit the free trade of these mineral commodities,' he wrote. By contrast 'one of the advantages of many of the mineral deposits of the sea is that they are generally equitably distributed throughout the oceans of the world and are available to most nations that might wish to mine them.'*[15] That led to the first golden age of exploration in the

---

* Mero believed that the sea would become a 'major source' of metals 'within the next generation.' 'Eventually, political and population pressures will force the more highly industrialized nations into recovering many materials from the sea,' he wrote. He went on: 'These ocean-floor sediments have other advantages when being considered as a material to mine, that of being politically-free and royalty-free materials, they

1970s, coinciding with a period when the world began to worry about running out of resources due to a growing global population and the effects of climate change.

However, one of the most prominent expeditions was actually an act of subterfuge. In 1974 the CIA carried out a plot to recover a sunken Soviet nuclear submarine under the guise of a fully planned deep sea mining expedition. At the height of the Cold War, in 1968, the K-129 Soviet diesel-powered sub had sunk in the deep waters of the Pacific, losing all of its crew. While the Soviet Union failed to find the sub, the US managed to narrow down its location. Former CIA director William Colby came up with a plan, which was code-named Project Azorian, that would give the Americans key insight into Soviet nuclear technology by lifting the 1,750-tonne sub from the seabed north-west of Hawaii. The plan took years to implement and involved respected researchers who went on to play a role in deep sea mining. The ship, the *Glomar Explorer*, was built with money from US billionaire Howard Hughes and used equipment developed by Lockheed Martin. It was launched at a champagne christening ceremony in 1972, along with a fake press release.[16] On 8 August two years later the submarine was lifted from the seabed with the help of an underwater vehicle fitted with a large claw. The CIA explained it like this in a tweet in 2019: 'Imagine standing atop Empire State Building w/8-foot-wide grappling hook on 1-inch-diameter steel rope. You must lower hook to street below, snag tiny car full of gold, & lift car back to top of building ...

---

are widely distributed near most markets and are available to all nations on an equal basis'; Mero, *The Mineral Resources of the Sea*, p. 275. He was also quite prophetic: 'Development of the means to mine the manganese nodules, thus, could serve to remove one of the historic causes of war between nations, supplies of raw materials for expanding populations. Of course it might produce the opposite effect also, that of fomenting inane squabbles over who owns which areas of the ocean floor and who is to collect the protection money from the mining companies.'

without anyone noticing.'[17] But on its way up, the submarine snapped in two, with the largest part falling back into the deep sea. Instead, the bodies of Soviet sailors as well as nuclear torpedoes were recovered. The Russian sailors were given a burial at sea. Despite its mixed success, however, the expedition helped inspire the first era of exploration of the seabed for minerals. In 1975 Lockheed leased the ship and started to explore the deep sea as part of Ocean Minerals Co., alongside Amoco, Shell Billiton and Boskalis.

It was one of a number of consortia formed in the 1970s from companies across the Western world. One of them, OMI, which included Japan's Sumitomo Group and Canada's nickel miner Inco, conducted the first successful pilot mining test in early 1978, pumping around 800 tonnes of manganese nodules from the floor of the Pacific. Two other consortia followed within a year. But just as soon as it had started, deep sea mining petered out and spending on expeditions was cut back significantly. 'In the mid-1980s, deep-sea mining for manganese nodules was dead in the water,' according to Ole Sparenberg.[18]

The main reason for the decline in interest was the fall in international commodity prices in the 1980s. But the other reason was encroaching regulation, which complicated efforts to mine for nodules. Mero had touted the advantages of deep sea minerals because they were not under the control of any government. But that was changing, as developing countries wanted to ensure any mining benefited them and that rich countries did not run off with the loot. In a speech to the United Nations General Assembly in 1967 the Maltese ambassador Arvid Pardo had proposed that the ocean floor be the 'common heritage of mankind'.[19] That led to the 1982 United Nations Convention on the Law of the Sea, which established governance over international waters outside of a country's coastal waters, which were defined as 200 nautical miles off the coast. Developing countries pushed for the establishment of a body to regulate deep sea mining, the International Seabed Authority (ISA), which finally came to fruition in 1994, based in Kingston, Jamaica. It was given the twin authority to come up with mining regulations but also to protect the environment and to manage mining activities 'for the benefit of mankind as a whole'.[20]

While mining in coastal waters, which fall under individual countries' jurisdiction, has taken place since then (such as for diamonds from the ocean floor off the coast of Namibia), no one has successfully started a commercial venture to mine the deep sea. Over the past decade, the ISA has given out fifteen-year exploration contracts to thirty countries or companies, covering more than 1.4 million square kilometres. Since the US did not ratify the Law of the Sea, Lockheed Martin operated in a partnership with the UK government and had two licences to explore 133,000 square kilometres in the Clarion-Clipperton Zone. China had the highest number of licences, with five, but the UK had the largest by area.

The ISA is led by Michael Lodge, a former barrister who spent many years living in the South Pacific. He had the unenviable job of getting 167 countries to agree in an age where the desire for global cooperation is teetering. A matter-of-fact Englishman, Lodge sees his mission as creating a 'solid regulatory framework' in line with the Law of the Sea convention and making sure it can be strictly enforced. He is putting in place rules for an activity that is already permitted, he tells me. Keenly aware of the emotional response that deep sea mining can provoke, he let little enter his own voice. When I spoke to him in early 2020 the final regulations had still not been concluded and countries were still debating how deep sea mining should be taxed. 'Nothing is agreed until everything is agreed,' Lodge said. The Law of the Sea stated that deep sea mining should not be given an unfair advantage over land-based mining, making it critical to come up with good data for comparison, a difficult task given the number of land-based mines and different regulatory regimes. Even when the rules were finally agreed by all member states Lodge cautioned that it would then take between eighteen months and two years for companies to make a formal application to actually mine the deep sea. 'It's not something that is going to happen tomorrow or immediately the regulations are adopted,' he told me. Some resource-rich African countries such as South Africa, a big manganese producer, wanted to make sure they would be compensated for any lost revenue from lower prices caused by supply of minerals from the deep sea, as the Law of the Sea had prescribed. Lodge said countries dependent on land-based mining would have to demonstrate 'on a case-by-case basis' the harmful economic impact of deep sea mining in order to be compensated. 'There's a number of bars to get over before you can establish that but the principle is already locked into the convention,' Lodge said. In any case, the royalties from deep sea mining would be distributed according to an 'equitable sharing formula' that favoured developing countries and small island states. Member states had converged on the idea of a progressive royalty, which is based on the value of minerals sold, which increases over time and 'captures any upside in metal prices whilst also protecting against the downside risks for the ISA', Lodge said. 'That would be in place until mining has actually started and we have an economic model based on reality rather than assumptions as we're working on now,' he added.

Lodge said he did not worry about the possible environmental impacts of deep sea mining, saying the methods for the environmental impact assessments and environmental monitoring programmes that would be required by companies were 'tried and tested' in offshore projects. 'There's an impact to every human activity in the ocean or on land for that matter, it's a human impact;

the question is how do we manage those impacts to minimise any damage that might take place,' he said.

On a cold February day in 2020 I took the Eurostar to visit GSR (Global Sea Mineral Resources), a subsidiary of Belgian company DEME, a dredging company that installs giant offshore wind turbines. I'm here to meet the Patania II, a robot designed to mine the deep sea that is named after the world's fastest caterpillar, the Patania ruralis. After a forty-minute drive across flat country from Antwerp to the port of Vlissingen, Kris Van Nijen, the managing director of GSR, stepped out of a smart black Mercedes to greet me before the chain-link gates to the dock. Surrounding us were some of the green-and-white ships with tall towers that installed the vast offshore wind turbines in the North Sea. He wore a dark blue polo neck and blazer and immediately exuded an intense energy. He told me he woke up at 5.45 that morning to do a forty-five-minute cycling workout – competing with his teammate, who is also called Kris. The other Kris, who had arrived in a separate vehicle, nodded in placid agreement with his boss. We put on hard hats, boots and fluorescent waistcoats and headed through the gates to a vast cold hangar, where Patania II sat: a 4.5-metre-high tank-like machine painted in DEME's signature bright green that looked like a cross between a bulldozer and a vacuum cleaner. It had rows of pipes at the front that suck up the nodules and sensors that can raise or lower the pipes to get the correct angle. 'It's like Robocop back in the day,' Van Nijen said, darting around the machine, which is reinforced like a military vehicle and is as tall as two people. It weighs twenty-five tonnes above water and fifteen tonnes below. The nodules are sucked up and then pushed to the back of the craft where they are stored. The sediment meanwhile is expelled out of giant vents at the back of the vehicle. Van Nijen took me to the back of the vehicle where the electronics were encased in thick titanium and bound together with large bolts, the only part that had to be kept dry at the immense pressures of the deep sea. Patania reminded me of a larger version of the autonomous vehicles used by the US military in Iraq to disarm bombs. The absence of any human on board was jarring, prompting me to think about the miners on land who will be replaced. Not that this robot would care.

Van Nijen told me the pressure is so great in the deep sea that there is no way to replicate it onshore in a tank. There is a one bar increase for every ten metres you go down, with the pressure reaching 451 bars, compared to one bar at sea level. 'It's like an elephant on a postal stamp,' he said. The vehicle will move across the ocean floor at 0.5 metres a second, he added. Van Nijen paced this out to me on the floor of the hangar. The other Kris showed me to a nearby white container inside which was a bank of computer monitors and joysticks to control the Patania II from the dock of the support ship.

Van Nijen had been a dredger all his life, and had only ever worked for DEME. The job had taken him all over the world, from Africa to Russia and four years in Singapore. Living in the Low Countries the Belgians knew all about flooding and dredging, he told me. He had worked for mining companies including diamond producer De Beers, helping it to suck up diamonds from the floor of the sea off the coast of Namibia. Before the financial crisis he had been approached by Barron's Nautilus Minerals to help them mine the deep sea but DEME decided against it. Still the idea stuck. In 2010 'we said, why don't we look at it rather than waiting for someone to come knock on our door,' he recalled. Van Nijen travelled the world for two years learning everything he could about deep sea mining and meeting everyone in the sector. 'It was easy then, there were only about 50 people,' he said. The more he researched polymetallic nodules, the more he realised it was just dredging, not mining, and DEME knew all about that! 'It is dredging technology,' he said emphatically. 'Not drilling, cutting, blasting, it is dredging – a simple vacuum system on the sea floor.' He paused. 'Only it's 5,000 metres deep; OK that's a problem,' he laughed. 'But we'll tackle that one.' The company decided that it was something they could do and bet $100 million on the project, putting him in charge. In 2013 GSR obtained a fifteen-year contract to explore in the Clarion-Clipperton Zone from ISA, with the support of the Belgian government. The company had taken 55,000 photos of their concession and mapped the nodules. 'There are millions!' Van Nijen said. 'Hundreds of millions of nodules, tonnes of nodules.' He wanted to make sure I got the point that it is *not* mining. 'It's dredging nodules; it's not mining nodules. You [can] call it harvesting nodules or dredging nodules, but it's not mining nodules,' he said, pointing at my notebook.

In early 2019 the Patania II sailed from San Diego for its first tests in the deep sea. It had undergone weeks of real-world testing, being lowered into the water by the dock, picking up lava stones in place of the deep sea nodules. Van Nijen did not really know what would happen when the machine attempted to collect the nodules from the seafloor. Would it sink into the thick layers of sediment that had accumulated over millions of years? Would the pressure crush its internal electronics? Minutes after it was launched into the middle of the Pacific from a large winch on board the ship, an alarm started ringing in the white-painted container where the younger Kris sat with his colleagues. They had lost all contact with the vessel. The mission was immediately over. There had been a problem with the 5,300-metre-long umbilical cord that connected the deep sea vessel to the ship's electricity system, which had led to a power surge. (The cord was under constant pressure from the pressure underwater but also from the ship's movement on the surface.) Patania II was winched back slowly onto the deck and Kris had to make a disheartening telephone call to CEO Kris.

Van Nijen told me he was disappointed but stoic – deep sea mining and exploration were hard, and if they weren't then everyone would be doing it. 'I guess all new industries have to deal with this,' he told me. The company even made a short film about the experience in an attempt to see the bright side. 'This movie is about failure,' Van Nijen said before showing it to me over lunch. 'You go to any business school and it's all about failure.'

NGOs, however, saw it a different way and greeted it as a sign that deep sea mining was doomed. As we drove in his new Audi Van Nijen sighed about the NGOs. 'Sometimes I just feel like giving up, I tell them OK I'll leave the planet to you and in ten years let's see where we are.' It all goes back to the Law of the Sea, he told me. 'How many of these NGOs have actually read the Law of the Sea?' he asked. The ISA rules could create a 'utopia' he said, a common resource managed by all of us, which was better than one country having the resource to themselves through the luck of geology. 'This might be the first example in human history of when an ecosystem has been the subject of such extensive research and the impact of operations assessed so thoroughly before any operations actually take place,' he said. To him the debate was frustratingly one-sided – you were either good or bad. 'The whole debate is so polarised,' he explained. 'Either you are good – you make Teslas and no one cares where the cobalt comes from, or you're the bad guy who mines the cobalt.' He believed the solution lay somewhere in the middle, involving little improvements – rather than step changes that come with absolutely zero downsides. EVs are not the huge improvements that people expected and wanted, he said. But they're still better than the internal combustion engine.

Deep sea mining would not be without any impact – but what would be the impact? Van Nijen believed that it should be based on facts, not emotions. If we don't know what's down there how can we be sure it will be damaged? he asked.

Van Nijen had been through all the academic studies and not one was predicting a decrease in demand for metals, he said, correctly. He criticised the NGOs who said deep sea mining would not shut down mines in the Congo – 'We're not trying to shut down existing mines but prevent new ones,' he said.

On his estimates, deep sea mining could eventually be a $1.5 billion-a-year business. 'That's why we invested $100 million into an industry that doesn't exist,' he told me. 'Where there is no regulatory framework. Who does that?' But Van Nijen said he was willing to give up deep sea mining if the science proved it would be damaging. 'If the research shows that the impacts are too high, that going to the seafloor for the minerals will not meet the world's demand for metal in a more environmentally responsible way than alternatives, then the industry will not proceed.'

While Van Nijen continued to prepare to mine the deep sea, another option to our mineral predicament was gaining steam: reusing the minerals we had already dug up.

13

Reduce, Re-use, Recycle: A Closed Loop

'Industry owes it to society to conserve materials in every possible way. Not only for the element of cost in the manufactured article, although that is important, but mostly for the conservation of those materials whose production and transportation are laying an increasing burden on society.'

Henry Ford, 1926[1]

J.B. Straubel had been obsessed with electric vehicles since he was a young man. As a teenager he once restored a golf cart he found at the dump, rebuilding its electric motor. While at Stanford he spent a year retrofitting an old Porsche 944 into an electric vehicle, using a lead-acid battery. The car could only drive twenty miles, so to make up for its range Straubel cut an old VW in half and attached the engine and rear wheels to the back of his car so it could push him along the road. Straubel drove the car 800 miles to Oregon. A photo taken in 2000 showed the machine at an electric car race. It could have been a contraption fashioned in Edison's day decades earlier – a bundled-together vehicle that promised a cleaner future but was far from being commercially feasible. At that time, to an outside observer who happened to be strolling along the street, it would have highlighted immediately how far electric cars were from being mainstream and how entrenched was the reign of the mighty internal combustion engine.

Yet Straubel's converted Porsche was lightning quick, setting the record for the quarter-mile at the Silent Thunder 2000 race of 17.278 seconds. It instilled in Straubel a sense of the potential of an electric vehicle. Soon he would be building much faster cars that would change the face of the entire global car industry.

In 2003, after he graduated from Stanford, Straubel received a visit from some old university friends at his home in Los Angeles. They had finished racing a solar car 2,300 miles from Chicago. That evening the friends hatched an idea to build a long-range EV, ditch the solar panels and drive on batteries alone. Instead of clunky lead-acid batteries, they would use lithium-ion. The batteries had improved significantly due to their use in consumer electronics. What would happen if you put thousands in a car?

It was not a question many were asking. In the US, the talk was all about the potential of hydrogen-powered fuel cell cars to wean the country off its oil addiction, not batteries. But Straubel soon found a fellow believer in Elon Musk. Later that year at a seafood restaurant in Los Angeles, Straubel met Musk, who had recently sold PayPal for $1.5 billion. When Straubel mentioned his efforts to build an electric car, Musk seized on the idea. Straubel subsequently became one of the first employees at Tesla, taking the title of chief technology officer at the age of thirty.

At Tesla, Straubel designed its first batteries, led development of the Gigafactory in Nevada and served as chief technology officer for fifteen years. So, when the soft-spoken engineer left in 2019, he could have raised money to do anything in Silicon Valley.

He decided to move permanently to Carson City, a town in the middle of Nevada. Once described by former resident Mark Twain as 'a desert, walled in by barren, snow-clad mountains', Carson City was home to fewer than 60,000 residents.[2] Straubel had bought a 300-acre ranch there to be near Tesla's Gigafactory.

He still drove an original bright yellow Roadster, one of the early prototypes of Tesla's first electric car. It was not the most obvious location for Straubel's latest start-up, Redwood Materials, whose mission was to break down discarded batteries and reconstitute them into a fresh supply of metals needed for new electric vehicles.

As Tesla had grown, Straubel had become aware of the quantity of raw materials that would be needed. He visited a nickel mine and was taken aback by its sheer size. 'It was becoming increasingly clear that we kept moving some of the challenges further upstream,' Straubel said.[3] Adding to the problem, battery factories such as the Gigafactory produced a large amount of waste materials from the battery manufacturing process – roughly around ten percent of material was wasted. That waste was worth hundreds of millions of dollars and could be re-used. If electric vehicles were to become mass-market items, Straubel realised, then recycling would have to be part of the story.

Redwood's warehouse was stacked high with boxes of discarded smartphones, power tools and old scooter batteries, which it received in truckloads from consumers and companies. Every weekday some sixty tonnes of waste arrived at the plant. 'Some things are just basically given to us, essentially for free,' Straubel said.[4] The batteries were heated in a furnace at 2,700°F to burn off the unwanted plastics and other organic materials such as binder, leaving a mix of metal powders. Redwood would then treat the metals with chemicals to reproduce materials such as cobalt sulphate and lithium carbonate, the building blocks for new lithium-ion batteries. Straubel recovered between ninety-five and ninety-eight percent of a battery's nickel and cobalt and more than eighty percent of its lithium. He believed there was a 'massive, untapped resource' of these metals in the garages of the average American. He estimated there were around one billion used batteries in US homes sitting in old laptops or mobile phones, all of which contained valuable metals. The number would only increase as people bought e-bikes and electric vehicles.

Metals, unlike fossil fuels, could be recycled endlessly, without any loss of function. It was a veritable gift from nature. The earth was full of metals that we could re-use once we had dug them up. Even once a lithium-ion battery was degraded after years of use, the same atoms of lithium, nickel and cobalt remained. They could be used to build new batteries. For the US, which lagged behind China in terms of access to metals such as cobalt, lithium and nickel, it made sense to start with what was already sitting at home in people's drawers. If you paid enough, you could get hold of the material. Straubel wanted to build a US supply base for battery materials to meet the growing demand from the country's largest carmakers such as GM, Ford and Tesla. 'How do you take something that exists, unmanufacture it and present it back into the supply chain so that a new thing can be made – and do that in a sustainable way that doesn't have other unintended consequences? That's what we're focused on,' he said.[5] His eventual ideal was for all transportation to be powered by electric vehicles, whose parts would be continually recycled so there was no need for mining. It would be a nearly closed-loop transportation and energy system. That would reduce the US's reliance on foreign raw materials and also help cut the cost of batteries, making electric cars even cheaper.

*

At the beginning of the car age there was little concern about the need to recycle raw materials. The early industrialists such as Henry Ford 'did not see their designs as part of a larger system, outside of an economic one'.[6] They relied on a 'seemingly endless' supply of natural resources, from ores to timber, water, coal and land. The United States was blessed with a rich abundance of what economists call natural capital. These raw materials were shipped to Ford's River Rouge factory in Dearborn, Michigan, and cars came out the other end. 'Resources seemed immeasurably vast,' Michael Braungart and William McDonough wrote in their seminal 2008 book *Cradle to Cradle: Remaking the Way We Make Things*.[7]

But Ford soon realised the value of re-using materials and the impact reducing waste would have on his costs. He became an early proponent of a 'lean manufacturing' system that would be later followed by Toyota to great effect in the 1990s. In the 1920s Ford started a 'disassembly line' to take apart used cars, with workers 'stripping each car of radiators, glass, tires, and upholstery as it moved down the line, until the steel body and chassis were dropped into an enormous baler'.[8]

It was an early step towards re-use of industrial materials. Since then, the car industry has been enormously successful: today around ninety percent of the steel in cars is recycled and over ninety percent of lead-acid batteries used in petrol cars are recycled. The use of recycled lead or scrap lead constitutes over half of the world's lead use, and over eighty percent in the US. The same is true for aluminium. When the first all-aluminium beverage can was introduced by Coors in 1959, the company offered one cent on every can returned, as the company could recycle the metal at a fraction of the cost of producing new aluminium. Today, a used beer can could be back on the shelf in sixty days.[9] It's estimated that three-quarters of the more than 1.4 billion tonnes of aluminium ever produced are still in productive use.[10]

Yet despite these improvements we are consuming more metals and minerals than ever before, a trend that the electric car and clean energy will only exacerbate. Between 2002 and 2015, global material extraction increased by fifty-three percent, according to one study. In those years over 1,000 gigatons of materials were extracted, or almost one-third of the total extraction since 1900, according to analysts at Bernstein. This is not slowing down: we're set to produce fifty percent more steel by mid-century than we do today, and plastics production is set to increase to 900 million tonnes from 300 million tonnes.[11]

Yet over ninety percent of the global economy uses natural resources 'unsustainably'.[12] While technological advances have allowed us to use materials more efficiently, such as replacing many gadgets with a smartphone, that has been outweighed by the growth in overall demand. It is a trend that was first noticed by British economist William Stanley Jevons in 1865, in the context of the UK's coal reserves. 'It is wholly a confusion of ideas to suppose that the economical use of fuel is equivalent to a diminished consumption. The very contrary is the truth.'[13] As a result of this, our emissions of greenhouse gases continue to rise. As the UK's chief scientist put it: 'Environmental challenges are not just about emissions. They are about resource consumption. Emissions are a symptom of rampant resource consumption. If we do not get resource consumption under control, we will not get emissions under control.'[14]

The effect can be seen clearly in the car industry: rules in the US that forced carmakers to improve their fuel economy over the last few decades have had no impact on overall demand for fuel, for example. Instead, people have just gone out and bought bigger cars and driven those cars even further. Vehicle miles travelled in the US have increased by one hundred percent since 1980.[15] In 1908 the kerb weight of the Ford Model T was 540 kilograms and three decades later the company's Model 74 weighed almost exactly twice as much.[16] Since 1990, US pick-up trucks have added around 1,300 pounds to their weight on average, with some vehicles now weighing 7,000 pounds, the equivalent of three Honda Civics.[17] All this weight means more raw materials.

Moving to what is known as a 'circular economy' by recycling and re-using is a way to make a dent in this continued demand for natural resources. It's a concept that has been embraced by some of the largest corporations from Apple to Dell. Yet much of it consists of little more than soundbites and public relations messaging. Most smartphones have a limited lifespan, in part because manufacturers such as Google, Samsung and Xiaomi only allow users to download security updates for a set length of time. In November 2020 Apple paid $113 million in the US to settle a lawsuit brought by over thirty states that accused it of secretly reducing performance in older phones, leading consumers to buy new ones. Our progress towards a circular economy has as a result been achingly slow: we throw away an estimated fifty million tonnes of electronic waste a year, less than twenty percent of which is recycled.[18]

Yet there is more gold in a tonne of this e-waste than in a tonne of mined ore, along with over thirty different raw materials including lithium and cobalt, tin and tungsten. The concentration of minerals in old electronics is much higher than it is in nature. The value of the waste is estimated to be at least $57 billion every year. An executive at Glencore told me we should stop calling it waste, but instead label it 'post-consumer scrap'.

Electric vehicles, with their larger batteries, will only add more waste, especially if we continue to buy larger cars such as SUVs. By 2025 there could be as much as six billion kilograms of lithium-ion batteries made every year, whose waste would be the size of the Great Pyramid of Giza.[19]

Recycling clearly represents a profound opportunity for the industry. But its impact will only ever be partial in the near term due to the growth in sales of electric cars. The best estimates suggest recycling of materials could provide up to twenty-five percent of the raw material inputs by 2025. Recycling is not the only way to get value out of existing batteries, however – re-use is also an option.

*

On an unseasonably cold and overcast day in late May, I decided to pay a visit to a shop in south London that re-used electric vehicle batteries to convert old cars. I had read about Ewan McGregor spending £30,000 retrofitting his 1954 VW Beetle to go electric. It seemed strangely counterintuitive: electric cars were the *new new thing*; how could they help revive an old thing?

Matthew Quitter, a friendly man with a greying beard and swept-back hair, opened a large wooden gate to welcome me to his garage under a railway arch in a narrow street near the Thames. Inside stood a number of old cars that had had their engines taken out and replaced with batteries and electric motors: a 1983 Land Rover, an old Mini and a proud black Bentley. The company mostly used Nissan Leaf batteries from cars that had been in crashes or been damaged. Quitter pointed me to a pile of them that were stacked up in the corner in their silver-metal cases. Most EV drivers had not been in serious crashes, he said. 'It's more the mid-level manager who's tired on a Thursday night and pulls out at a cross-section and gets T-boned,' he said. The batteries were also well protected inside a large black steel casing. Electric cars that had done 100,000 miles may have lost some of their capacity, he said, but they were still good enough for London driving. 'Most people drive predictable routes – they go from home to somewhere else where the car sits; then perhaps somewhere else where it sits again, then home,' he told me. 'It's a common misconception that they [batteries] would no longer be viable,' he said. 'They are perfectly good batteries. We believe after the second life they will have a third life, then a fourth and a fifth.' Quitter got hold of old batteries at fifty percent of their cost even though they still had eighty to ninety percent of their capacity left.

Quitter started his business after he turned his Morris Minor classic car into an electric by fitting it with a battery and electric motor. He had developed an interest in classic cars after returning to London from New York, where he worked in music writing commercial songs in a 'jingle factory' for clients such as Club Med after studying at Brown University. Half-American, Quitter had quit music to work in digital publishing but soon tired of the business. His Morris Minor conversion managed to garner some press and he soon realised there was a market for the product, so in 2016 he left his job and set up on his own.

Quitter didn't want to only convert cars for rich enthusiasts, however. 'That's just high-end hot rods,' he said, referring to cars that are fitted with bigger engines. He wanted to lower the cost of conversions so that his company could make a meaningful difference to carbon emissions. 'The car was seen as a symbol of freedom because people could get in and go an indeterminate distance,' Quitter told me. 'But the cost of it is that everyone gets one and that is killing the planet.' He felt it was a waste for the UK to scrap so many of its cars so that people could buy new electrics. He told me there was a disused airfield north of London where thousands of old cars were just sitting there, waiting to be scrapped, some of them perfectly good cars. There were fourteen million cars on the road in the UK, and the government wanted to ban sales of petrol and diesel cars by 2030. 'Half of a car's emissions happen during its construction. There's really a strong argument to have the opportunity to convert people's cars to electric. We think the government should be endorsing this,' Quitter said.

After we talked in his office, which was full of old parts from Nissan Leaf electric cars, including speakers, battery management systems and control systems, Quitter took me for a ride in the

Land Rover. At first the car did not start; there seemed to be some problem with the smaller twelve-volt battery. 'This is not starting out well, is it?' he joked. Quitter looked intensely at a little dial that had been appended to the front dashboard which gave the state of capacity of the battery. Then after a bit of tweaking with the key, the car started and Quitter shifted into third gear. We were off. 'There's no use for this clutch of course,' he said, pushing it down. We headed out of the garage's gate and onto the London streets. It was a robust, old Land Rover, but it was eerily quiet. 'The fact we're having this conversation and not trying to shout over the noise is an improvement,' Quitter said as he turned onto a busier street. 'Some people think this is sacrilegious,' he added a few moments later, and I asked why. He told me that many English people believed that a car had to be kept in the same pristine condition it was in when it was produced, whereas in the US there was a much more liberal attitude to conversions. Yet most of his customers were attached to their old cars, they had been in the family for years, and they wanted to keep them in the face of tightening regulations on emissions.

Quitter's work highlighted the need to make products last longer. Much of the short lifespan of consumer electronic goods was due to the practice of 'planned obsolescence', which began in the 1920s when General Electric and other lightbulb makers created a cartel to shorten the lifespan of lightbulbs, so people bought more of them. It worked remarkably well. Today, the same incentive remains in place: it generally costs more to repair an item than to buy a new one. Electric vehicle batteries, however, were likely to last much longer than people thought.

Hans Melin, a recycling expert based in Muswell Hill in London, spent his days tracking EVs and trying to work out where older models were ending up. He found many of them were sent out of their home markets for sale in developing countries – just like petrol cars. Between 2015 and 2018, an estimated fourteen million older vehicles were exported from Europe, Japan and the US, to developing nations, according to the United Nations.[20] Melin tracked sales of electric vehicles in developing nations and found that Ukraine was a particularly strong buyer of used EVs. The top five selling electric vehicles in Ukraine in January 2021 were mostly all used vehicles from the US and the European Union, he found. 'In Ukraine you can find a lot of Chevy Bolts and a lot of cars from the American market,' he said. But Melin also found that electric cars were being used for much longer than people expected. The average age of a car in the UK was eleven years, while in the Ukraine it was about eighteen years. As a result, Melin predicted that by 2030 only sixteen percent of the total used lithium-ion batteries available for recycling would be available in Europe and ten percent in the US. Instead of exporting polluting old cars to the developing world, we were now exporting clean energy. 'We used to export pollution and now we are exporting zero emissions – that's hard to complain about,' he said.

Even when a battery has been run down so much that it is not suitable for an electric vehicle, it could still be re-used in other applications such as for storage of renewable energy or to make other forms of transportation electric, such as boats. One study found that EVs with eighty percent residual battery capacity could still meet the daily travel requirements of eighty-five percent of people in the US.[21] Our batteries will probably outlive our cars (and, in the future, possibly us). Recycling and re-use would not solve all of the challenges of battery production. But combined with greater consciousness about the energy supply involved in building batteries it heralded a more optimistic approach. The question was who would champion recycling and building greener batteries on an industrial scale.

14

The World's Greenest Battery

'The battery will be the key differentiator in the car. People
will care a lot about the battery and how it was produced.'
Financial backer of European battery start-up Northvolt

In early 2021 large rectangular low-lying buildings started to
emerge out of the surrounding snow in the north of Sweden, 125
miles south of the Arctic Circle. The perfectly smooth snow-cov-
ered roofs that broke the cover of the surrounding trees looked like
giant non-descript warehouses, or the site of a secret military science project. Yet the town of Skellefteå felt their presence acutely:
it was undergoing its biggest boom since gold was discovered in
the 1920s, with the hotel rooms all fully booked. Six hundred
construction workers, of all nationalities, had descended on the
town, where they lived in a small village especially built for them.
Work continued throughout the Covid-19 pandemic, which
stranded many Polish workers who could not return home. 'It's
probably the largest hotel north of Stockholm,' Peter Carlsson,
Northvolt's founder, told me. Skellefteå was now set to play a key
role in powering Europe's electric vehicle revolution, producing
homegrown batteries. The 600-metre-long white buildings, which
represented only a quarter of the eventual size of the battery factory, were a geopolitical statement of ambition: Europe was in the
global battery race and it was deadly serious.

In 2012 Peter Carlsson was working day and night at Tesla in California in preparation for the launch of the company's first luxury electric car, the Model S. A protégé of Elon Musk, Carlsson's job was to manage the car company's global supply chain, dealing with over 300 suppliers. A tall Swede with closely cropped hair, Carlsson had built up his expertise as head of supply chain at Sony Ericsson in Sweden and NXP Semiconductors in Singapore. In 2000 at Sony Ericsson, Carlsson was in charge of buying around $1 billion worth of batteries for their mobile phones, from Sony and Panasonic. (Yasuo Anno, the general manager of Sony Energy at the time, would eventually join Northvolt.) Then Carlsson had dropped it all to join a young start-up in Silicon Valley that was nowhere near making any money. Working at Tesla was fast-paced and gave Carlsson a deep understanding of the difficulties of scaling up manufacturing of electric cars.

At Tesla, Carlsson saw that the next ten years would be an 'industrial land grab' in the global battery industry. 'The one who is most successful in attracting the best people and scaling up production and bringing financing will be a winner,' he said. After quitting Tesla at the end of 2015, he began to think about the need for Europe to have a homegrown battery champion. He returned to Sweden and teamed up with another Tesla alumnus, Paolo Cerruti, whom Carlsson had first convinced to join Tesla from Renault-Nissan in Paris. In the fall of 2016, they decided to form Northvolt, operating initially under the name SGF Energy. They studied the feasibility of building a battery Gigafactory in Europe and spent six months travelling the world to meet customers, suppliers and politicians. They started to recruit battery experts. Carlsson soon settled on a goal: to build the greenest battery in the world using Sweden's ample supplies of renewable energy. It was a bold challenge to the Asian giants who dominated battery manufacturing: Panasonic, LG Chem, Samsung SDI and China's CATL. 'I never thought we were working in an environmentally friendly industry,' Yasuo Anno, who joined the team, said. 'So Peter's plan is really meeting that kind of requirement – also my dream.'[1] Carlsson knew from the beginning that scale would be the key factor. 'This is really an industry where you either go big or you go home,' he said. 'If nobody does anything, Europe is going to be completely dependent on an Asian supply chain.'

*

Europe had sleepwalked into the electric vehicle age. While Tesla developed its first electric car, European car companies were promoting diesel as the solution to reducing emissions, since burning it produces less carbon dioxide than petrol. It was an idea that died a quick death following revelations in 2015 that Volkswagen had cheated on its emissions tests by installing a 'defeat device' in its diesel engines that could detect when they were being tested. By 2016 the European car industry was at a crossroads: carmakers faced the prospect of billions of euros in fines for breaching mandatory limits of carbon dioxide by 2020, yet diesel was out of favour.[2] The only path to meet the regulations was to go electric. Yet apart from a few producers, such as Saft (which was bought by oil giant Total), Leclanché and France's Bolloré, Europe produced no batteries on the scale needed. 'If we don't take back innovation, it will be like the solar industry all over again,' Anil Srivastava, chief executive of Swiss battery producer Leclanché, told me, referring to the loss of Europe's solar panel manufacturing industry a few years earlier. Yet making batteries was a capital-intensive industry that few European carmakers had expertise in. 'Does anyone in Europe want to get involved?' analysts at the American brokerage firm Bernstein asked sceptically.

It turned out Brussels did. Officials were realising that the continent was woefully unprepared for a shift to electric vehicles and risked shifting a reliance on the Middle East to one on China. Forecasts suggested there could be forty-four million electric cars on the roads of Europe by the end of the decade. The European car industry employed one in every twenty workers.[3] It was an unpalatable idea to rely on China, Korea and Japan for the technology needed for the industry to survive. In addition, relying on foreign battery suppliers meant that a large share of the value of an electric car was provided outside the EU. In late 2017 the EU's vice president Maroš Šefčovič, a former Slovak diplomat who had run unsuccessfully for president of his home country in 2019, decided to turn the tide. Well-built with wire-framed glasses, Šefčovič had noticed an increasing number of Chinese electric buses in Brussels.[4] In October he formed the European Battery Alliance, a body charged with catalysing financing for the battery industry via the European Investment Bank. 'Our objective for the Alliance is simple, but the challenge is immense,' he said. 'We want, almost from scratch, to create a competitive and sustainable, battery cell manufacturing in Europe supported by a full EU-based value chain.'[5] Europe needed at least fifteen to twenty-five Gigafactories to catch up with Asia, Šefčovič said – requiring around twenty billion euros of investment.

Šefčovič saw the issue in starkly geopolitical terms. China was out to challenge Europe's automotive sector and upend the post-war multilateral world order. China was also moving up the value chain while 'locking users into new dependencies', via its securing of critical minerals such as cobalt and lithium. 'We cannot sit idle while China is taking control of all the supply,' he said in a 2019 speech. Europe produced only fourteen percent of the world's nickel, eight percent of its cobalt and one percent of its lithium and graphite. There was also no refining or processing of these elements in Europe – meaning materials would still need to be shipped to China. Europe's goal was 'strategic autonomy', Šefčovič said.[6]

Europe decided to follow China's example and heavily subsidise a homegrown battery industry. It was an acknowledgement of the need for an industrial policy to compete in the markets of the future. Europe had spent years complaining about China's subsidies to its homegrown companies; now it decided to copy the practice. Šefčovič understood that competing with China solely on price would not work: he believed that producing greener batteries that had a lower environmental footprint would be key to giving Europe an edge. Europe would pioneer 'sustainable battery production', he said, which would include 'extraction with the highest environmental and ethical standards, production with the lowest carbon footprint possible ... battery re-use and recyclability of materials'.[7] According to Volkswagen the battery was responsible for over forty percent of the emissions produced from making an electric car. The carbon dioxide emissions from making a battery in China were around sixty percent higher than in Europe due to the country's reliance on coal, according to researchers at Bloomberg New Energy Finance. Yet Volkswagen had pledged to become carbon neutral across its entire supply chain. If it relied on China for batteries, this would be hard, if not impossible, to achieve.

Northvolt offered a solution. Sweden had one of the greenest electricity grids in Europe and the company's factory would use one hundred percent renewable energy. Northvolt also pledged to make its batteries using up to fifty percent of recycled materials by 2030. In addition, the company pledged to have a supply chain that was 'free from conflict, child labour, and human rights abuse'.[8] Making batteries consumed an enormous amount of energy: Northvolt's factory in the north of Sweden would account for close to 2.5% of the country's electricity consumption once it reached its full capacity of 60 gigawatt hours of battery production. 'We know it's energy intensive and done right it's for the good but if that creates a large carbon footprint it kind of misses a little bit of the point of the transformation,' Carlsson told me. In 2018 Northvolt secured a €52 million loan from the EU's European Investment Bank to build a pilot battery line in Sweden. The following year the bank promised to provide a €350 million loan for Northvolt to build its battery plant in the north of Sweden. That support helped bring in private investors and, in the summer of 2019, Northvolt secured $1 billion led by Volkswagen and Goldman Sachs. Goldman Sachs became a key backer, taking three seats on the company's board and investing a further $1 billion in the summer of 2021.[9] Northvolt was now raising substantial sums of money and showing that it could compete with China's capital-raising abilities. 'It's very important that a number of battery factories are built in Europe,' Carlsson said later. 'Because there is no way that just LG and Samsung can supply the entire market.'

By 2021 Northvolt had to deliver on its promises. In March, half a year after Tesla's 'Battery Day' event, Volkswagen launched its own 'Power Day', which was broadcast online from a stage in Germany, using bright green letters that were reminiscent of a 1980s sci-fi film. 'E-mobility has won the race,' VW's CEO Herbert Diess, who was dressed in a dark suit and white shirt with no tie, said. 'It is the only solution to reduce mobility emissions fast. Our goal is to secure a pole position in the global scaling of batteries.' The car manufacturer said it needed six battery cell factories in Europe, and had agreed to pay Northvolt $14 billion for battery cells. It was Northvolt's biggest order. 'Our problem is not the demand, it is how to manage to execute on all these requests,' Carlsson told me and my colleague a few months later, as the company's battery factory neared its first production date.

Europe had arrived. In 2020, Europe's sales of electric cars surpassed China's for the first time. It marked a significant shift in the centre of the global EV market. For years China had been the leader. Yet in 2019 European investment in electric transport – at €60 billion – was more than three times higher than China, which invested €17.1 billion, according to Brussels-based non-profit Transport and Environment. Just a year earlier Chinese investment was seven times what Europe was investing in the sector. Europe's rapid growth sparked concerns in China that Beijing was not doing enough to support the electric vehicle market. In a speech in late 2019 CATL's chairman Robin Zeng called on China to introduce stricter policies similar to Europe's mandatory carbon emissions limits. 'If in the coming few years there is still this trend, if there's no investment there won't be production, and it will be very hard for us to continue to stay in the first echelon,' Zeng said.[10] Šefčovič was happy: he claimed that by 2025 the EU would be self-sufficient for battery cells. Of the 272 battery Gigafactories in the pipeline globally, twenty-seven were now in Europe, according to Benchmark Mineral Intelligence. It was an acknowledgement that reaping the jobs and economic growth from batteries and reducing reliance on China would require government support.

*

The success of Northvolt was an important step in competing with China, and a move that put Europe much further ahead than the US. Yet there were not enough European mines to expand this growing supply chain that the company needed. Most of Europe's nickel would come from Russia's Norilsk Nickel, a company controlled by the Russian oligarch Vladimir Potanin, and its lithium from Chile or Australia. It was an example of how Europe would still be reliant on other countries for raw materials – reducing its goal of strategic autonomy, and negating some of the environmental benefits of producing its own batteries.

In the summer of 2020 over 21,000 tonnes of diesel oil leaked into the Ambarnaya River from a fuel tank at a power plant near the Russian Arctic city of Norilsk, turning the waters red. The spill was so severe that Russia's President, Vladimir Putin, called a state of emergency. Greenpeace compared it to the 1989 Exxon Valdez disaster in Alaska. 'The scale of the damage to Arctic waterways is unprecedented,' Russia's ecology minister, Dmitry Kobylkin, said. The impact of the diesel spill was likely to last for years, and spread through the water and soil into the food chain, affecting local animals and birds.[11] A group of academics estimated it could take decades to clean up. Russia's environmental watchdog ordered Norilsk to pay a record $2.1 billion fine over the spill.

In August 2019 a network of indigenous groups in the Arctic wrote an open letter to Elon Musk demanding he avoid buying Norilsk's nickel until it re-cultivated contaminated lands, compensated locals for environmental damage from all its activities and promised to seek their approval for future exploitation. The Aborigen Forum said pollution was a 'routine occurrence'. 'The lands of indigenous people appropriated by the company for industrial production now resemble a lunar landscape,' they said. 'And traditional use of these lands is no longer possible.'*[12]

The town of Norilsk lies north of the Arctic Circle, where temperatures regularly drop below −10°C. Carpeted with snow for eight months of the year, the city is one of the most isolated and polluted places on the planet. It has its roots as a former Gulag town under Joseph Stalin, where hundreds of thousands of prisoners were forced 320 kilometres deep to build and operate Soviet mines. Though there is no official death toll it's thought that 100,000 people were killed. Today around 180,000 people live in the city, surrounded by ageing Soviet-era smelters that are some of the biggest emitters of sulphur dioxide, which helps cause acid rain and deforestation. In 2018 Norilsk emitted 1.9 million tonnes of sulphur, compared

---

* A month later Norilsk released a statement saying it had developed a five-year support programme for the indigenous people living in the Taimyr peninsula, pledging two billion roubles ($26 million). This money would be for protecting the natural habitat, supporting traditional activities, and for housing, medical, 'infrastructural, touristic, educational, and cultural projects'.[13]

to a total of 84,000 tonnes emitted by global mining giant Rio Tinto. It has also a history of environmental accidents: in 2016 the Daldykan River in Norilsk turned bright red, forcing the company to admit that heavy rain had caused a filtration dam to flood into the river. Yet the mine Norilsk owns contained the world's richest deposits of nickel, palladium and platinum.

A large portion of the wealth of Norilsk goes to one man: Vladimir Potanin, Russia's richest man, who owns 34.5% of Norilsk Nickel. In late 2018 I met Potanin in a private room known as the 'Boardroom' in the quiet surroundings of the mezzanine floor of Claridge's hotel. At times Potanin, with his smooth bald head and jowly face, could look like a Bond villain. But he was also disarmingly relaxed and smooth. Potanin told me about how the company was looking to enter the electric car supply chain, and had just announced a deal with Germany's BASF to produce battery materials together in Finland. 'For us it's an opportunity to go deeper into the value chain to see how it works, to have exposure not only on the nickel and cobalt and other metals production but also certain exposure on the development of electromobility,' he said.[14] The deal would help Norilsk clean up its environmental footprint and become more sustainable, he pledged. Potanin's colleague, marketing director Anton Berlin, was more forthright in a later interview to media: 'Batteries are the best thing that has happened to nickel in a hundred years, since the invention of stainless steel,' he said.[15]

Potanin had got his stake in Norilsk via the controversial 'loans for shares' scheme he helped instigate under President Boris Yeltsin four years after the collapse of the Soviet Union. Yeltsin handed over stakes in some of Russia's most valuable natural assets to a group of Moscow businessmen, in return for loans to the state. The businessmen then ended up selling the shares essentially to themselves. Potanin got a thirty-eight percent stake in Norilsk for just $170.1 million, which was worth around $15 billion in 2020.

In early 2022, before the outbreak of Russia's war in Ukraine, the company was worth $42 billion.

Norilsk was too big to ignore. The company said it could supply enough nickel by 2030 to produce between 3.5 and 5.5 million electric car battery packs, which could reduce global carbon dioxide emissions by 50 to 100 million tonnes. The deal with BASF inserted Norilsk right into the heart of Europe's EV supply chain – tying the European Union ever closer to Putin's Russia, just as it tried to escape from its reliance on Russian natural gas. It is another example of how the minerals we need for the energy transition are changing geopolitics – and tying countries together.

Even with Northvolt's move to control battery production, it was hard to see how Europe could be completely free of the darker elements of the materials supply chain. One entrepreneur in the UK, however, had a bold idea that could be part of the solution.

15

Cornwall's Mining Revival

On an overgrown bramble- and bracken-covered path on the hills above St Austell Jeremy Wrathall walked towards a shallow pit that still bore the marks of human activity from over half a century earlier. He picked up a twenty-centimetre-wide rock from the floor, pointing out the small flakes of lithium-rich mica that glinted a soft brown in the Cornish midday sun. 'We don't know how deep it was, how much they took out,' Wrathall, a former investment banker who quit his job to hunt for lithium in Cornwall, said. 'The collective memory is gone. It's like geological detective work.' Wrathall was tipped off about the UK's only known historical lithium mine, which was active in the Second World War, by a phone call from a retired member of the British Geological Survey and found it with the help of satellite images 'and some detective work'. He believed the mine helped supply lithium to help remove carbon dioxide in submarines, at a time when Germany was Europe's biggest producer of the mineral.

We are surrounded by small conical hills, which are old waste piles – the result of mining for 'China clay', or kaolin, which was used to make porcelain for centuries. The hills are known as the

'Cornish Alps', Wrathall's young colleague told me. They loomed over the nearby town of St Austell, where we could see the China clay plant owned by the French company Imerys, which sat adjacent to rows of neat small houses. The area looked eerily empty, the opposite of the factories I was used to seeing in China, with their uniformed workers and smokestacks. Wrathall told me the industry had been severely affected by the reduced demand for paper, which used the grey-white clay. It was another example of how Cornwall, one of the most deprived areas of the UK, had found it increasingly hard to compete in the global economy. In the 2016 referendum Cornwall had voted overwhelmingly to leave the European Union. But Wrathall thought that lithium offered the chance of hope. Cornwall had been a mining centre for over four thousand years, supplying tin and copper – a period captured by Winston Graham's *Poldark* series of novels set in the eighteenth century, which had recently been made into a BBC series. But globalisation and development of new sources of metal supply had left Cornwall with not a single active mine. An entire industry had been decimated, with mine sites reduced to historical relics for tourists and film crews. Wrathall believed Cornwall had abundant natural potential that had been overlooked – enough sun to generate solar power, strong Atlantic winds for offshore wind turbines, and hot geothermal fluid flowing deep under the ground that could generate energy and heat and also provide lithium. Could Cornwall be part of an alternative to Chile or Australia as a lithium producer using its abundant renewable energy? 'It's the ultimate Brexit success story,' Wrathall said.

*

A broad-shouldered man with a gentle manner, Wrathall grew up in the Lake District where his grandfather had founded a

Mountain Rescue service. At the age of three his family moved to Woking, which he didn't much like. 'Woking isn't much of a place,' he told me. A restless boy, it wasn't clear what Wrathall would become as a teenager. His father told him: 'You like to blow things up, you like digging things up and engineering – what are we going to do with you?' His father advised him to study mining and apply to the Camborne School of Mines in Cornwall, the UK's premier place for geology and mining, which promised a life of adventure and digging stuff up. 'It was the best decision of my life,' he recalled.

After graduation Wrathall moved to South Africa, which was then under apartheid, to work on a deep gold mine for a division of miner Anglo American. Managing a local team of South Africans in the deep hot mine was 'the closest thing to a vision of hell you can get', he recalled. He had to learn a special language used in the mine called 'fanakalo', which was a pidgin language that could be understood by all workers in the mine. One time, while working miles below ground, a rock fell on him and cut his shoulders and arms. 'If it was a heavier rock I could have nearly died,' he recalled. Another time there was a fire underground. 'I really thought that was me done for,' he said. After a few years he had had enough. He was about to become a father so decided to return to England to go into the City. He joined the traditional City broker Cazenove, where he was 'a bit of an experiment', he said. 'I was the only one who hadn't been to public school.' He caught the end of London's traditional City broking boom – a time of boozy lunches and expense accounts. Wrathall then moved on to UBS and then Deutsche Bank, where even more money was thrown around – 'They paid for everything, there would be fully paid trips to Marbella.' In the 1990s London was the place to be.

But in 2001 Wrathall decided to give it all up and return to the Lake District to join a mountain rescue team like his grandfather, and open an outdoor clothes shop. For this keen mountaineer, it was a chance to get out of London and reconnect to where he was from. He would work in the shop with an alert system that was pinged when people needed rescuing – walkers who had got lost, or those who went out woefully unprepared for a change in the weather. But the idealistic vision of escaping London quickly gave way to reality, and it became a rough time. He used to lie awake worrying about how he would sell all his stock in his shop. In the end he needed to start again.

When a friend asked him if he wanted to join a company doing copper exploration in Zambia he jumped at the chance. He became a specialist in African mining, travelling to the Democratic Republic of the Congo to work on a cobalt and copper deposit, where he was forced to sleep in a tent out in the bush. In 2011, as China's commodity demand accelerated, Wrathall thought about re-opening Cornwall's old Wheal Jane tin mine, which had previously been owned by Consolidated Goldfields and then Rio Tinto in the 1960s and 1970s. Metal prices were high and everyone was rushing to open mines wherever they could. The mine was on land owned by Lord Falmouth of the Boscawen family, who had owned the Tregothnan estate since the thirteenth century. In 2012 Wrathall had an expensive lunch with the heir to the estate. It was a contact that was to prove very useful for him. He gave up the idea of opening the mine, however, and after a few more finance jobs in 2013 he joined Investec as a mining analyst. He would travel the world to visit mines. In September 2015 he became famous for writing a note saying that Glencore's equity was worthless, causing the stock to fall by thirty percent, wiping £2 billion off the company in the biggest fall of any FTSE 100 company ever from an analyst note.

By 2016, however, the China-led commodity boom was over and life as an analyst was becoming increasingly constrained by regulation. At the same time sales of electric cars were taking off, and the mining industry was beginning to focus on the raw materials needed to decarbonise the world economy. Coal mines were out and lithium was in. In February 2016 Wrathall was walking to work in the City when he remembered a remark from a friend who had worked for the South Crofty tin mine in Cornwall, which had shut in 1998 with the loss of 200 jobs. The friend had said there was lithium in the hot salty fluid that flowed through the rock at the bottom of the mine. He had given him some maps a few years earlier, which Wrathall had put on a storage unit and forgotten about. It was a 'dark and horrible rainy day', he recalled. 'And I thought, really, is this what life is all about?' He got home and googled lithium in Cornwall, finding that it was not just in South Crofty but all over Cornwall, contained in hot saline water at the bottom of mines. The UK could be sitting on a lithium mine in its own backyard! He decided to quit his job and start Cornish Lithium, giving up the comfortable salary of Investec and becoming an entrepreneur. His colleagues 'thought I was crazy ... but the job of an analyst is pretty awful – it wasn't a job I enjoyed anymore.' He paused. 'It was probably the most insane choice I've ever made,' he recalled.

*

In early March 2020, just before Covid-19 struck the UK, I took the train down to Cornwall with Wrathall. The media was full of stories about decoupling from China and Chinese supply chains – the UK was struggling even to make the most basic personal protection equipment itself. The regional airline Flybe had just gone bust, severing Cornwall from air traffic, and forcing us to change our plans and go by train. I was pleased – I much preferred the slowness of the train, and it felt like a foretaste of what was to come as airlines would soon be forced to ground all flights. We sat at a table by the window in the afternoon sun. As we crossed the River Tamar and left Devon Wrathall joked, 'You're in foreign country now', to indicate we had crossed the natural border into Cornwall. After we passed through one tunnel the view suddenly opened out and we trundled along next to the sea, which felt like a final release from the city as we watched the waves cresting white in the wind. Wrathall was keen to show me the comments by UK minister Nadhim Zahawi the previous evening at a dinner to mark the launch of a new Critical Minerals Association, in which Zahawi had supported the development of UK resources. 'The potential to become self-sufficient in lithium which Cornish mining represents will I think be incredibly important to the British economy,' he had said.[1] Boris Johnson, the prime minister, had also been supportive when he was asked the previous week in Parliament about lithium in Cornwall, saying, 'It is a wonderful thing that Cornwall indeed boasts extensive resources of lithium, and we mean to exploit them.'[2] Other politicians, however, were less enthusiastic. When asked in Parliament about Cornwall's lithium by the local MP Steve Double, Jacob Rees-Mogg had replied haughtily that the UK had relied on fair and free trade for its industries and that would continue.

After founding Cornish Lithium Wrathall had started a 'voyage of discovery like you couldn't believe' into Cornwall's mining history, searching for mentions of lithium in old newspaper clippings and trying to get hold of old maps and documents. One day he drove to Essex to visit a friend who had got in touch saying he had an archive at home. He visited the British Geological Society, which had big dusty boxes of archives in a basement room, and offered to digitise them in return for a small fee. He went to visit the Coal Authority, who also had lots of old records, as well as old copies of *The Mining Journal* going back to 1835. Bit by bit, he started to create a digital archive of Cornish mining history, looking for mentions of lithium. He found that the historical records talked about miners stripped to the waist due to the heat produced from springs at the bottom of the mines, which limited how deep they could mine. In 1864 these hot springs were shown to have high concentrations of lithium by a London professor at King's College London, W.A. Miller. Miller examined a sample of hot water taken from a thermal spring at the bottom of the Wheal Clifford mine and noted that it may be valuable due to the use of lithium in the medical industry as a mood stabiliser at the time. 'The occurrence of so large an amount of lithium, being eight or ten times as much per gallon as has been found in any spring hitherto analysed, invests this water with unusual interest and importance,' he wrote. 'And as lithium has been employed medicinally to an extent hitherto limited by its high price, it may prove of great commercial value.' All told Wrathall found seventy-seven mentions of lithium in the historical record and digitised a total of twenty terabytes of data from over a thousand old maps.

At the company's offices in a modern building near the Camborne School of Mines Wrathall showed me the map of Miller's discovery of lithium in United Mines, entitled 'Transverse Section on the Hot Lode'. At the bottom right-hand corner was a neatly drawn box saying: 'The temperature of the deep levels in this mine is remarkable,' noting that it was 110° at 220 fathoms and 124° at 230 fathoms. 'The hot water issuing in great quantities at these depths is rich in Lithia,' it said, using the old term for lithium. Wrathall asked me to lean in close to the map, where at the top of the mine shaft we can see two small stick men fighting. 'The drawer had a sense of humour,' Wrathall added. I was then allowed into the closely guarded archive room, by Neil Williams, who grew up in Cornwall and used to work for the county archives. Williams told me he still remembered the explosions from the South Crofty mine (now closed) at 3 p.m. every afternoon when he came back from school – 'no one batted an eyelid,' he recalled. Old mining maps were stacked from floor to ceiling, along with notebooks and cassette tapes given to the company from the archives of Jack Trounson, an early twentieth-century expert on Cornish mining who had taken many photos and conducted interviews with Cornish miners. 'He died of a broken heart because the Cornish mining industry had broken down,' Williams told me. He pulled out one old table-sized mining map of the Wheal Busy mine, which showed in minute detail the wheel houses and shafts going deep into the earth, and the mineral faults that cut across the maps in short red lines. The detail of the maps was remarkable, he said, given they only had a candle to see by when underground.

The data from the old maps was put into software that allowed Cornish Lithium to create a 3D model of the earth's crust, giving them a critical clue on where to drill to intercept the fissures containing lithium-rich hot water. The company had traced the geological structures all the way off the coast of Cornwall into the seabed. 'It's really powerful being able to combine it into one place. Obviously, they could put two maps next to each other but they couldn't spin them around and look at it in 3D and visualise it,' Lucy Crane, a senior geologist for Cornish Lithium who graduated from the Camborne School of Mines, told me. 'I've been impressed by how accurate they were.'

Having found the historical documents, Wrathall next had to secure the rights to look for lithium in Cornwall. Many mineral rights were still owned by the big landed estates, but others had been sold on many times, and were separate from land rights. Like the shale oil pioneers in the US, Wrathall found himself visiting the local landed gentry, and signing agreements to be able to explore on their mineral rights. By 2021 Wrathall had rights to explore large areas for lithium as well as in areas under licence from the Crown Estate off the north and south coasts of the region.

*

Mining in Cornwall has taken place since at least the early Bronze Age, around 2100 to 1500 BCE, and there is evidence of trade between Cornwall and the Mediterranean by the fourth century BCE. Since as early as 1500 the extraction of tin continued almost uninterrupted until 1998. 'There is little evidence that any of the great events of history – such as the invasion by the Romans and their withdrawal 350 years later – did any more than temporarily disturb that international trade,' according to one history.[3] The area once had as many as 3,000 tin and copper mines during its heyday in the nineteenth century, when prices for copper rose because of the Napoleonic Wars and the use of copper to protect the hulls of Royal Navy ships. Gwennap became the richest square mile in the world due to the wealth from the copper mines. But by the 1870s copper prices had collapsed and the miners as well as the capital emigrated to newer areas in America, Canada and Australia in the great Cornish migration. 'All the men and all the capital left,' Wrathall said. Even today in Mexico there is an English Cemetery – Panteón Inglés de Mineral del Monte – that contains the graves of Cornish miners who flocked to the mineral-rich land in the nineteenth century.[4] 'By the late nineteenth century there was barely a mine in the world that did not have Cornish labour, and many had Cornish mine captains,' as historian Sharron Schwartz wrote.[5] Tin mining continued, however, and in the 1960s and 1970s there was a brief phase of renewed exploration in Cornwall, carried out by Consolidated Goldfields and Rio Tinto. But a collapse in the tin price in 1985 decimated the industry and the last Cornish mine, the South Crofty mine, closed in 1998 with the loss of around 200 jobs. Tin production shifted to Myanmar, Malaysia and Indonesia. Today the biggest producer of tin is Indonesia, a country that is almost completely powered by coal, where there are few environmental controls on mining. 'The whole history was completely forgotten about,' Wrathall said. 'Everyone gave up on Cornwall – all the men left, and the capital left. It's re-evaluating an old mining area with modern eyes.'

Cornwall sits atop a 280-million-year-old slab of granite, which is covered by sediment and extrudes through the earth's crust at places like Dartmoor, 'like a Loch Ness monster', according to Lucy Crane, the young geologist with Cornish Lithium. Through fissures in the rock, or faults, water had percolated down into the very heart of the granite, absorbing and dissolving lithium during its journey. These were the hot springs that Cornish miners found at the bottom of their mines. The old maps tracked these courses, and Wrathall's company could now plot them in 3D – tracing them across Cornwall and all the way off the coast into the Atlantic. The Cornubian Batholith, as the granite is known geologically, also contains radioactive elements such as uranium and thorium that give off heat as they decay, making it perfect for geothermal energy. 'It's like a giant nuclear reactor deep in the earth,' Wrathall said.

If you could extract lithium using geothermal energy and heat it would offer a low carbon way of mining, reducing the need for an external power source. Efforts to extract lithium from geothermal resources were beginning to bear fruit in the Salton Sea in California, a man-made lake where geothermal power was already produced for the grid. The advantage of so-called 'direct lithium extraction technologies' was that they didn't require the evaporation of brine in large ponds such as in Chile. Instead, the hot geothermal brine was passed through filters that extracted the lithium, after it had been used to turn a turbine to generate power. Then the brine was returned into the earth. As Alex Grant put it, 'geothermal lithium projects could kill two birds with one stone: producing low CO_2 intensity lithium chemicals for lithium-ion battery manufacturing and decarbonizing electricity grids simultaneously.'[6]

*

The next day we drove out from Falmouth along narrow roads bordered with tightly trimmed hedges towards the drilling site, where Wrathall's team were looking for lithium with the help of data from the old mining maps. Wrathall had been nervous about what the company might find. 'It was a huge risk, possibly an insane risk,' he said. Traces of the history of mining seemed to be everywhere: we passed old derelict mining engine houses, where steam engines had pumped water up from the mines. Then a pub that used to make pit props for the mining industry. A river that used to be a port. Wrathall told me about how someone's porch had collapsed because of an old mining tunnel beneath. At one closed tin mine, the Wheal Jane mine, which Wrathall had once thought of re-opening, we could see large fields of tailings waste still laid out in the sun, a thick crust of dark mud and toxic metals. After the mine closed in 1991 a total of over ten million gallons of heavily contaminated water burst from the mine and swept into the Carnon River, spreading all the way to Falmouth Bay.[7] A processing plant built by Veolia was still dealing with the waste water years later. It was a reminder of the toll of mining, which felt strange to see in England, and not in some far-flung country.

At the drill site a group of men from Ireland manned a yellow-painted diamond drill rig in the cold wind, every so often pushing perfectly round pieces of drill core out from steel pipes onto a flat surface. The drill cores were examined and logged by a staff member in a nearby container office, where a computer model of the rock included the data from the old mining maps. Looking like grey marble, the core of the rock was dotted with feldspar and quartz, and flashed gold in colour in some places. 'That's fool's gold,' Lucy Crane told me. More exciting were the bits of lithium-rich mica. Crane told me she had given up a career in Africa or Latin America to stay in the UK. 'You get rose-tinted glasses hiking around the high Atlases in Morocco but it's nice to be able to balance a really interesting career and having a life. There's so much potential in the UK for geo-resources, it just needs to be developed responsibly, the social licence to operate will be really sensitive,' she said. Some of the core had crumbled like broken tofu, a sign of fractures in the granite through which hot lithium-rich water had flowed.

The question – which Wrathall set out to answer – was whether Cornwall's resources were big enough to sustain a competitive industry on the scale needed to meet our demand for batteries. Cornish Lithium would need to demonstrate a large, low-cost resource in Cornwall's granite to gain funding, Grant, the expert in the sector, told me. 'The UK needs to be honest about whether their lithium resources are globally relevant or not,' he said. 'Economy of scale will still win, and VW is not going to give you an offtake [of lithium] unless you have a consistent vetted product.'

*

I wondered also about who would really benefit from the resumption of mining in Cornwall. Land ownership was starkly unequal, with the Prince of Wales and landed estates owning most of the mineral rights. On my last day we visited the Tregothnan estate, a large Tudor Gothic building sitting above lush gardens on a hill near Falmouth. The Estate's agent, Andrew Jarvis, met us outside the house due to Covid-19 and walked us around the botanical gardens, which he said were first cultivated in 1264. 'This bush is over 250 years old,' he said at one point. A sharp and shrewd man, who used to work for the steel industry, Jarvis now looked after the family's investments, which spanned everything from farming to technology start-ups. On a raised piece of land, he asked me to stand on a ledge and look down at the tea plantations the estate grows. 'We have a microclimate that means it is the same climate as Darjeeling,' he said. The estate produced some of the only commercial tea grown in the UK, which they sell in Fortnum & Mason and Claridge's hotel. Next Jarvis showed me a rare tree that was thought to have been extinct before an Australian scientist rediscovered it. The Lord had stayed up all night on an online auction to buy the tree, a Wollemi pine, and have it shipped back to Cornwall, he said. As we walked on, he briefly recounted the family history – how they had made money from mining, gambling, horse racing and the Navy. In the eighteenth century Edward Boscawen, an admiral in the Royal Navy, had captured some French ships in Canada, and taken a large amount of prize money. 'It can be considered a good day at the office,' Jarvis said. We then returned to the main house, where the Boscawen Rose flag flew high above the buildings. I asked him about lithium and he brightened immediately. 'It's a continuity with what we've been doing for centuries,' he said. 'It is a gamechanger for Cornwall if it's managed properly and it could be for the UK as well. We can bring mineral extraction home to the UK.'

But other landowners in Cornwall were more cautious. One landowner, whose family has owned his estate since the thirteenth century, told me he was hesitant about a rush of new mining ventures, even though he had signed an agreement with Wrathall. The family had seen all the peaks and troughs of Cornwall's mining industry and his estate still had the derelict remains of old tin mines. We needed to remember that not everyone had got rich, and that more money was lost in mining than was probably made, he said. The average lifespan of a miner had been thirty-five years – ten years less than an agricultural worker for only a little more money a week. 'It was a dirty, nasty industry ... There's a disjoint in how it was in *Poldark* and what it was actually like,' he told me. 'Although it was of great economic benefit to Cornwall, a lot of people lost out. There's a lot of love of mining but a realisation that this is a transitory thing. We want to avoid boom and bust – we want managed economic growth rather than a smash and grab.' A gentle man with a patrician air, he saw himself and the other landed gentry as 'custodians of the landscape'.

It was a challenge normally faced by countries in Africa or Latin America. But if the UK really wanted to reduce its reliance on China and develop its own supply chains, the country needed to rise to it. Projects like Wrathall's in Cornwall would not solve the whole problem, but they could help spur on new mines and initiatives across Europe.

Conclusion

After almost two hundred years we are coming to the end of the age of oil. Our children are already growing up with a different understanding of what the car means. On a sunny day in July of 2017, I travelled with my wife to Hull, a town in the north of England, to see high schoolers race electric cars they had built themselves. The goal of the race was to go as far as possible on one battery charge. It was not about speed but efficiency of engineering. The children I spoke to understood intuitively why this mattered. The ideas of scarcity, preservation and climate change were top of their minds. Where the previous generation might have focused on the thrill of speed and taken for granted the abundance of oil, these children were already planning for an electric future. It was a hopeful message.

Once known for its shipping and fishing industries, Hull was changing in other ways too. At the end of 2016 a large factory outside the town produced its first wind turbine blades. The 250-foot-long blades were exhibited in the city centre as the biggest thing ever made in Hull. Off the coast the world's biggest offshore wind farm was being built. A battery factory in Sunderland, which had been bought by China's renewable energy company Envision, was pumping out ever more batteries for electric vehicles and energy storage. 'We're at a tipping point,' Stephen Irish, a youthful-looking engineer who had set up a battery company outside the plant, told me in November 2017, as he drove home in his electric BMW. He had written his thesis on electric cars in 1996, and had held hope all those years. After years of delay the transition to clean energy was happening everywhere I looked.

The transition to electric cars, renewable energy and batteries will create a greener, better world. Electric cars not only improve health outcomes but can also boost economic growth and bolster productivity – especially in China. According to a study examining eighteen years of pollution records in China, people fled from polluted cities, especially skilled, educated employees. In the UK, one study found that the impact of pollution from roads impacts more than seventy percent of the country's land area. 'Potentially less than 6% of land escapes any impact, resulting in nearly ubiquitously elevated pollution levels,' the authors wrote.[1]

But we need to clean up the invisible supply chains we have come to rely on. Consumer goods, even green ones like electric cars, create what the scholar Peter Dauvergne called 'ecological shadows', which are shifted onto poorer countries least able to cope with them. We need to be fully aware of what ecological shadows will be created by the green revolution. Dauvergne's words are prescient: 'Not only is environmentalism failing to produce sustainable patterns of global consumption, much of what policy makers in high consuming economies are labelling as "environmental progress" is in reality little more than the wealthy world deflecting consequences and risks into ecosystems and onto people with less power – and thus less influence over global affairs.'[2]

We can choose to reduce our appetite for SUVs and large cars. If we continue to choose large vehicles, we will be choosing large, heavy battery packs and increasing the burden on mining. The number of SUVs has grown from less than 50 million in 2010 to over 280 million. More than fifty-five percent of the electric vehicle models announced worldwide in 2020 were SUVs and pick-up trucks, according to the International Energy Agency. This is because they are more profitable for carmakers, so the extra margin outweighs the costs of transitioning to electric vehicles. But continuing on this trend will put pressure on the planet. We could instead prioritise the use of large batteries to electrify the vehicles that are on the road every day: trucks, vans, lorries. They operate much more than any of us drive our cars.

We also need to make battery production and mining more environmentally friendly – a process that can begin once we acknowledge and recognise the issues. Mining trucks can become electric or powered by hydrogen fuel cells. We can make refineries and battery factories run on renewable energy. We can also use renewable energy to make hydrogen to produce steel. As the sustainability scholar Auke Hoekstra puts it: 'One could imagine a future in which not only the cars themselves but the entire automotive supply chain runs on renewable electricity. Batteries could power the mining equipment that retrieves the ore from which batteries, solar panels, and windmills are made. Solar and wind produce hydrogen that (in combination with batteries) makes the production of steel and aluminium almost zero emission, which in turn makes the manufacturing of batteries, cars, solar panels, and wind turbines almost zero emission. Car batteries can also absorb energy from solar and wind power, helping to stabilize the grid, and in turn reducing the amount of stationary batteries that are needed.'[3] Such a world is possible and offers us all hope.

Technical innovations will reduce the burden on some resources. BYD, China's largest electric carmaker, has mastered a type of lithium-ion technology that had been neglected by battery producers outside China for many years. Known as lithium iron phosphate (LFP) batteries, they contain no nickel or cobalt, but only lithium and iron, two abundant elements. LFP batteries are less powerful than those containing nickel and cobalt, meaning they are unlikely to power longer-range cars, pick-up trucks or lorries. But they are good enough for driving shorter distances. In 2021 the bestselling electric car in China wasn't a Tesla but a tiny Wuling Mini with an LFP battery that cost around $4,500. The car could travel 100 miles on a single charge.

We need to be continually aware, however, that ecological shadows are always shifting. New battery technologies in development might reduce demand for one raw material, only to replace it with another with an equally convoluted supply chain. Increasing demand for raw materials will put greater pressure on the natural environment, already suffering from loss of biodiversity and the impacts of climate change. Growth based on extraction cannot be infinite. As Vaclav Smil has written: 'Continuous material growth, based on ever greater extraction of the Earth's inorganic and organic resources and on increased degradation of the biosphere's finite stocks and services, is impossible.'[4]

Changes in the supply chain will only come with continued consumer pressure. Public awareness forces companies to reform how they produce materials, and secure their energy. Chinese companies, as in the case of Huayou Cobalt and Tsingshan discussed in this book, react very quickly when their customers want them to. We need to become active consumers and, as Christiana Figueres put it, 'vote with your money' for products that are sustainable.[5]

Any electricity-based transportation system will not come to us from the gods, but will rely on extraction of the world's precious resources.

Acknowledgements

Thanks to Simon Moores for all his help and support over the years, and also to Caspar Rawles and Andrew Miller at Benchmark Mineral Intelligence for all their insight and help. This book, however, does not reflect the views of Benchmark Mineral Intelligence.

Thanks to Wang Xiaoshen for explaining the lithium industry to me and letting me visit Ganfeng in China, and also to Tianqi Lithium for the site visit and interviews. Joe Lowry has always been generous in sharing his knowledge and expertise, as have John Kanellitsas and Jon Evans. Alex Grant was always helpful on lithium extraction and carbon footprints. Ken Brinsden has always been willing to talk and insightful about the state of the lithium market. In Chile, thanks to SQM for the help and site visit, and also to Alonso Barros, Eduardo Bitrán and Óscar Landerretche. Thanks to Robert Friedland for all the conversations and for letting me visit the Kamoa mine in the DRC, and also to Alex Pickard and Matthieu Bos at Ivanhoe for arranging my visit and letting me stay at the mine site. Thanks to Stan Whittingham, Ram Manthiram, Billy Wu and Bob Galyen for talking to me about batteries. Hans Eric Melin has always been helpful over the years on battery recycling. At CATL thank you to Elaine Huang for her help. At Glencore thanks to Charles Watenphul for arranging the visit to Mutanda and for help over the years. Elisabeth Caesens was always generous with her time on all things DRC, as was Nicholas Garrett. Jeremy Wrathall was extremely generous with his time and showed me around Cornwall, just before the first Covid-19 lockdown. For expert advice and feedback thank you to Joe Lowry, Anneke Van Woudenberg, Hans Melin, Steven Brown, Jim Lennon and Adrian Glover. Any mistakes are all mine of course. There are others I can't name who have helped me understand battery materials and I thank them. Apologies if I've left anyone out.

Thanks to Geoff Dyer, Tom O'Sullivan and Murray Withers at the *Financial Times* for their help with the *FT* Big Reads, where I first covered many of these issues. Thanks to Tom Wilson for all his insight into the DRC, and Patrick McGee for his work on battery recycling; to the whole commodities team – Neil Hume, Anjli Raval, Emiko Terazono and David Sheppard – for their support and camaraderie; and to Katie Martin and James Lamont for letting me take on this project.

At Oneworld thanks to Alex Christofi for first taking this book on and to Cecilia Stein for shepherding it through to completion. Thanks also to Kathleen McCully for the careful copy-edit. My agent David McCormick believed in this topic early on, when electric cars and batteries were not in the public eye, and instantly grasped what I wanted to say.

Finally, thanks to my friends Matt and Ryan, my parents, and Siri and Jo Harris for reading the book. And most of all to my wife Claudia for all her insight and advice, and to our little son Jamie who can't get enough of our electric car.

Notes

Introduction

1   Tesla Q4 2019 earnings conference call transcript, 29 January 2020, Motley Fool, www.fool.com/earnings/call-transcripts/2020/01/30/tesla-inc-tsla-q4-2019-earnings-call-transcript.aspx (all URLs were checked on 27 January 2022).

2   Timperley, J., 'How our daily travel harms the planet', *BBC Future*, 18 March 2020, www.bbc.com/future/article/20200317-climate-change-cut-carbon-emissions-from-your-commute.

3   McKibben, B., 'If the world ran on sun, it wouldn't fight over oil', *Guardian*, 18 September 2019, www.theguardian.com/commentisfree/2019/sep/18/climate-crisis-oil-war-iraq-saudi-attack-green-energy.

4   Fisher, A., 'Tony Fadell's next act? Taking on Silicon Valley – from Paris', *Wired*, 19 October 2017, www.wired.com/story/tony-fadell-revenge-on-silicon-valley-from-paris/.

Chapter 1   The Battery Age

1   Tesla Battery Day presentation, YouTube, 22 September 2020, www.youtube.com/watch?v=l6T9xIeZTds.

2   Ibid.

3   Ibid.

4   Watts, S., *The People's Tycoon, Henry Ford and the American Century* (New York, Vintage, 2009), p. 288.

Chapter 2   Dashed Hopes: The Troubled History of the EV

1   Strohl, D., 'Ford, Edison and the cheap EV that almost was', *Wired*, 18 June 2010, www.wired.com/2010/06/henry-ford-thomas-edison-ev/.

2   'Edison batteries for new Ford cars', *New York Times*, 11 January 1914.
3   'World mobility at the end of the twentieth century and its sustainability', Massachusetts Institute of Technology and Charles River Associated Incorporated, October 2001, https://docs.wbcsd.org/2001/12/ Mobility2001_FullReport.pdf.
4   Morris, E., *Edison* (New York, Penguin Random House, 2019), p. 273.
5   Crawford, M., *Why We Drive: On Freedom, Risk and Taking Back Control* (London, Bodley Head, 2020), p. 31.
6   McCall, B., 'My life in cars', *New Yorker*, 12 December 2020.
7   Watts, *The People's Tycoon*, p. 46.
8   Ibid., p. 42.
9   Skrabec, Q., *The Green Vision of Henry Ford and George Washington Carver: Two Collaborators in the Cause of Clean Industry* (Jefferson NC, McFarland & Co., 2013), p. 91.
10   '1883 interview with Thomas Edison on energy storage', *Seeking Alpha*, 9 December 2013, https://seekingalpha.com/instablog/227454-john-petersen/2479201-1883-interview-with-thomas-edison-on-energy-storage. Extracted from the *New York Sunday Herald*, 28 January 1882.
11   Morris, *Edison*, p. 219.
12   Ibid., p. 220.
13   Bakker, G., 'Infrastructure killed the electric car', *Nature Energy*, 6, 947–8 (2021).
14   Kirsch, D., *The Electric Car and the Burden of History* (New Brunswick, Rutgers University Press, 2000), p. 31.
15   Morris, *Edison*, p. 236.
16   Morris, *Edison*, p. 249.
17   'Edison's new battery, as applied to automobiles is now on exhibition', *New York Times*, 18 January 1903.
18   Josephson, M., *Edison* (New York, McGraw-Hill, 1959), p. 415.
19   Mom, G., 'Inventing the miracle battery: Thomas Edison and the electric vehicle', in G. Hollister-Short, ed., *History of Technology*, vol. 20 (London, Bloomsbury Publishing, 2016), p. 32.
20   Kirsch, *The Electric Car and the Burden of History*, p. 200.
21   Ibid., p. 13.
22   Kirsch, *The Electric Car and the Burden of History*, p. 200.

Chapter 3   The Breakthrough: The Lithium-Ion Revolution

1   Stanley Whittingham banquet speech at Nobel Prize ceremony, 10 December 2019, www.nobelprize.org/prizes/chemistry/2019/whittingham/speech/.
2   Belli, B., 'Nobel-winning alum John Goodenough '44 B.A. inspires next generation of Yale scientists', *Yale News*, 9 October 2019, https://news.

yale.edu/2019/10/09/nobel-laureate-john-goodenough-44-inspires-next-generation-scientists.

3   Interview with Stanley Whittingham, November 2019, www.volkswagenag.com/en/news/stories/2019/11/we-must-believe-in-the-unbelievable.html.

4   Gardiner, B., *Choked: The Age of Air Pollution and the Fight for a Cleaner Future* (London, Granta, 2019), p. 147.

5   Goodenough, J.B., *Witness to Grace* (PublishAmerica, 2008), p. 51.

6   Rhodes, R., *Energy, A Human History* (New York, Simon & Schuster, 2018), p. 172.

7   'Ford demonstrates an electric auto; seeks new battery', *New York Times*, 20 July 1967.

8   Banerjee, N., Cushman, J., Hasemyer, D., Song, L., *Exxon: The Road Not Taken* (InsideClimate News, 2015), p. 2.

9   Goodenough, *Witness to Grace*, p. 17.

10   Manthiram, A., 'A reflection on lithium-ion battery cathode chemistry', *Nature Communications*, 11, 1550 (2020), https://doi.org/10.1038/s41467-020-15355-0.

11   Manthiram, A., Goodenough, J.B., 'Layered lithium cobalt oxide cathodes', *Nature Energy*, 6, 323 (2021).

12   Goodenough, *Witness to Grace*, p. 73.

13   Fletcher, S., *Bottled Lightning: Superbatteries, Electric Cars, and the New Lithium Economy* (New York, Hill & Wang, 2011), p. 56.

14   Yoshino, A., 'From polyacetylene to carbonaceous anodes', *Nature Energy*, 6, 449 (2021).

15   Wald, M., 'Improving batteries', *New York Times*, 14 June 1989.

16   Lewis, L., 'Japan powers up for the new battle over batteries', *Financial Times*, 30 March 2021.

17   Trancik, J., Ziegler, M., 'Re-examining rates of lithium-ion battery technology improvement and cost decline', *Energy and Environmental Science*, 14 (2021), 1635–51.

Chapter 4   China's Battery King

1   Robin Zeng, email to CATL employees in 2017. Captured from CATL official WeChat account, but since deleted.

2   Chazan, G., 'Tesla's Berlin plant sets up "duel" with German carmakers', *Financial Times*, 15 November 2019.

3   Scott, A., 'Can Europe be a contender in electric-vehicle batteries?', *Chemical & Engineering News*, 13 July 2020, https://cen.acs.org/energy/energy-storage-/Europe-contender-electric-vehicle-batteries/98/i27.

4   'Mercedes-Benz and CATL as a major supplier team up for leadership in future battery technology', *PR Newswire*, 5 August 2020, www.prnewswire.

com/in/news-releases/mercedes-benz-and-catl-as-a-major-supplier-team-up-for-leadership-in-future-battery-technology-831440057.html.

5   'Interview with Prof. Friedrich Prinz', Volkswagen, www.volkswagenag.com/en/news/stories/2019/05/e-mobility-transition-interview-with-professor-prinz.html.

6   Details of Zeng's life are from Chinese media articles.

7   Fletcher, *Bottled Lightning*, p. 59.

8   锂业课堂 | 解析'电动中国':中国锂电池产业的崛起之路，任道重远, Tianqi Lithium official WeChat channel, https://mp.weixin.qq.com/s/WGP81G1G3X3uRX_w2btVtQ.

9   Ibid.

10 李曙光, '宁德时代, 不能永远只做选择题', *Tencent News*, https://new.qq.com/omn/20200302/20200302A03D5B00.html.

11 See family memorial blog: 'Memorial to Tang-Hua (TH) Chen, PhD', https://tanghua-chen.muchloved.com/Gallery/Thoughts/7766885.

12 See 'Carlyle leads a US$30 million investment with 3i as a co-investor in a leading Chinese battery maker, Amperex Technology Limited', Carlyle press release, 17 June 2003, www.carlyle.com/media-room/news-release-archive/carlyle-leads-us30-million-investment-3i-co-investor-leading.

13 Ibid.

14 Ibid.

15 Tao, W., 'Recharging China's electric vehicle policy' (Carnegie-Tsinghua Center for Global Policy, 2013), https://carnegieendowment.org/files/china_electric_vehicles.pdf?TB_iframe=true.

16 Eckhouse, B., 'The US has a fleet of 300 electric buses. China has 421,000', *Bloomberg News*, 15 May 2019.

17 Mazzocco, I., 'Electrifying: how China built an EV industry in a decade', *Macro Polo*, 8 July 2020, https://macropolo.org/analysis/china-electric-vehicle-ev-industry/.

18 'Next generation lithium: from electronic components to the smart grid of the future', *Batteries International*, Issue 92 (summer 2014).

19 Li, F., 'Powerful CATL dominates electric car battery sector', *China Daily*, 11 March 2019.

20 'The breakneck rise of China's colossus of electric-car batteries', *Bloomberg Businessweek*, 1 February 2018.

21 Schreffler, R., 'Battery supplier CATL riding crest of EV wave', *Wards Auto*, 29 January 2018, www.wardsauto.com/technology/battery-supplier-catl-riding-crest-ev-wave.

22 Ibid.

23 Ward, A., 'Low-carbon technology power play by China's CATL', *Financial Times*, 15 March 2018.

24 'CATL aims to plug into the global market', *China Daily*, 29 December 2016.

25 Li, F., 'Powerful CATL dominates electric car battery sector', *China Daily*, 11 March 2019.

26 'Battery pack prices cited below $100/kWh for the first time in 2020, while market average sits at $137/kWh', *Bloomberg New Energy Finance*, 16 December 2020, https://about.bnef.com/blog/battery-pack-prices-cited-below-100-kwh-for-the-first-time-in-2020-while-market-average-sits-at-137-kwh/.

27 Conference call with Bernstein analysts, April 2020.

28 Zeng, Y., '抓住重大历史机遇，推动我国新能源产业快速发展', GaoGong Industry Institute, www.gg-lb.com/art-40852.html.

29 Ibid.

Chapter 5 The Chinese Lithium Rush

1 Elon Musk Twitter account, 20 July 2020.

2 Fioretti, J., Zhu, J., 'Ganfeng Lithium in dreadful Hong Kong debut, may bode ill for rival Tianqi', *Reuters*, 11 October 2018, www.reuters.com/article/ganfeng-lithium-listing-idUSL4N1WR17E.

3 'Lithium comes from exploding stars', NASA, 29 May 2020, www.nasa.gov/feature/lithium-comes-from-exploding-stars.

4 Fles, A., 'Should we all take a little bit of lithium?', *New York Times*, 14 September 2014.

5 See Brown, W., *Lithium: A Doctor, a Drug, and a Breakthrough* (New York, Liveright, 2019).

6 Comer, E.P., 'The lithium industry today', *Energy*, June 1978.

7 Kinzley, J., *Natural Resources and the New Frontier, Constructing China's Borderlands* (Chicago, University of Chicago Press, 2018), p. 150.

8 Coughlin, W., 'Into the outback', *Stanford Magazine*, March/April 2000, https://stanfordmag.org/contents/into-the-outback.

9 'The lithium-ion supply chain', *Benchmark Mineral Intelligence*, September 2016, https://s1.q4cdn.com/337451660/files/doc_articles/2016/161214-Benchmark-approved-for-distribution-Lithium-ion-supply-chain.pdf.

10 Ingram, T., 'Pilbara Minerals boss Ken Brinsden's "fortuitous" leap from iron ore to lithium', *Australian Financial Review*, 4 November 2017.

11 Goldman Sachs research note, referred to here: www.goldmansachs.com/insights/pages/what-if-i-told-you-full/?playlist=0&video=0.

12 Burton, M., 'Australian lithium recovery seen by mid-2020 as EV production revs up', *Reuters*, 18 September 2019.

13 Grant, A., Deak, D., Pell, R., 'The CO_2 impact of the 2020s' battery quality lithium hydroxide supply chain', Jade Cove Partners, https://jadecove.com/research/liohco2impact.

14   Raby, G., *China's Grand Strategy and Australia's Future in the New Global Order* (Melbourne, Melbourne University Publishing, 2020), p. 2.

15   Sanderson, H., 'Australia seeks investment from European electric carmakers', *Financial Times*, 17 September 2019.

Chapter 6   Chile's Buried Treasure

1   Dubiński, J., 'Sustainable development of mining mineral resources', *Journal of Sustainable Mining*, 12, 1 (2013), 1–6.

2   'Eduardo Bitrán, vicepresidente ejecutivo de Corfo: "Ponce Lerou representa una época muy desgraciada de nuestra historia"', *The Clinic*, 22 November 2015, www.theclinic.cl/2015/11/22/eduardo-bitran-vicepresidente-ejecutivo-de-corfo-ponce-lerou-representa-una-epoca-muy-desgraciada-de-nuestra-historia/.

3   De Onis, J., 'Allende accuses US copper interests', *New York Times*, 12 July 1971.

4   Méndez, P.G., *The Reinvention of the Saltpeter Industry* (SQM Department of Communications, Santiago, Chile, December 2018), p. 90. Author's personal copy.

5   Ibid., p. 90.

6   Herrera, H., 'el ex Sindicalista de SQM que busca hundir a Ponce Lerou', *The Clinic*, 12 April 2015, www.theclinic.cl/2015/04/12/hugo-herrera-el-ex-sindicalista-de-sqm-que-busca-hundir-ponce-lerou/.

7   See 'Take the money and run? The consequences of controversial privatizations', www.lse.ac.uk/lacc/publications/PDFs/Gonzalez-Privatizations.pdf.

8   O'Brien, T.L., Rohter, L., 'The Pinochet money trail', *New York Times*, 12 December 2004.

9   Details of this history come from Zhang, X.H., 'The secret history of Tianqi lithium industry', *Economic Observer*, 10 June 2020, http://m.eeo.com.cn/2010/0610/172275.shtml.

10   Ibid.

11   Sanderson, H., 'China warns Chile against blocking $5bn SQM lithium deal', *Financial Times*, 26 April 2018.

Chapter 7   The Cobalt Problem

1   Moores, S., 'Has Glencore given electric vehicles the extra push to engineer cobalt out of a battery?', *Benchmark Mineral Intelligence*, 16 November 2018, www.benchmarkminerals.com/has-glencore-given-electric-vehicles-the-extra-push-to-engineer-cobalt-out-of-a-battery/.

2   *The Economist*, 17–23 February 2018.

Chapter 8　The Rise of a Cobalt Giant

1　Thomas Jefferson to Horatio G. Spafford, 17 March 1814, https://founders. archives.gov/documents/Jefferson/03-07-02-0167.

2　Glencore Nikkelverk, 'Sustainability', www.nikkelverk.no/en/sustainability.

3　Van Vuuren, H., *Apartheid, a Tale of Profit, Guns and Money* (Johannesburg, Jacana Media (Pty) Ltd, 2017), p. 111.

4　Breiding, J., 'Yes, he played dirty – but Marc Rich also changed the world', *Financial Times*, 27 June 2013.

5　United States Congressional Serial Set, No. 14778, House Report No. 454.

6　'DR Congo stands to lose $3.71 billion in mining deals with Dan Gertler', *Raid*, 12 May 2021, www.raid-uk.org/blog/ drc-congo-stands-lose-3-71-billion-mining-deals-dan-gertler.

7　See Melman, Y., Carmel, A., 'Diamond in the rough', *Haaretz*, 24 March 2005.

8　Ibid.

9　Ibid.

10　Richburg, K.B., 'Mobutu: a rich man in poor standing', *Washington Post*, 3 October 1991, www.washingtonpost.com/archive/politics/1991/10/03/ mobutu-a-rich-man-in-poor-standing/49e66628-3149-47b8-827f-159dff8ac1cd/.

11　Stearns, J., *Dancing in the Glory of Monsters: The Collapse of the Congo and the Great War of Africa* (New York, PublicAffairs, 2012), p. 165.

12　US Embassy Cable, Wikileaks, https://wikileaks.org/plusd/ cables/01KINSHASA1610_a.html.

13　Wilson, T., 'DRC president Joseph Kabila defends Glencore and former partner Gertler', *Financial Times*, 10 December 2018.

14　'Cutting-edge diplomacy', *Africa Confidential*, 24 October 2003, www. africa-confidential.com/article-preview/id/189/Cutting-edge_diplomacy.

15　Melman and Carmel, 'Diamond in the rough'.

16　'The mineral industry of Congo (Kinshasa)', United States Geological Survey, 1997, https://minerals.usgs.gov/minerals/pubs/country/1997/9244097.pdf.

17　Stearns, *Dancing in the Glory of Monsters*, p. 289.

18　Ibid., p. 319.

19　Heaps, T., 'Tea with the FT: young blood', *Financial Times*, 7 April 2006.

20　Mahtani, D., 'Transparency fears lead to review of Congo contracts', *Financial Times*, 3 January 2007.

21　'Deciphering the $440 million discount for Glencore's DR Congo mines', *Resource Matters*, November 2017, https://resourcematters.org/wp-content/ uploads/2017/11/Resource-Matters-The-440-million-discount-2017-11-29-FINAL-1.pdf.

22 'Fleurette and Glencore complete merger of Mutanda and Kansuki mining operations', *PR Newswire*, 25 July 2013, www.prnewswire.com/news-releases/fleurette-and-glencore-complete-merger-of-mutanda-and-kansuki-mining-operations-216882021.html.

23 'Glencore and the gatekeeper, how the world's largest commodities trader made a friend of Congo's president $67 million richer', *Global Witness*, May 2014, www.globalwitness.org/en/archive/glencore-and-gatekeeper/.

24 Hearn, A., 'EXCLUSIVE: Tycoon pays £46m for London flat (plus £3m more in stamp duty)', *Evening Standard*, 7 January 2015.

25 Oz Africa Plea Agreement and Statement of Facts, 29 September 2016, www.justice.gov/opa/pr/och-ziff-capital-management-admits-role-africa-bribery-conspiracies-and-agrees-pay-213. Dan Gertler is not mentioned by name in the document, but the description of the DRC Partner matches him.

26 Rose-Smith, I., 'Dan Och's African nightmare', *Institutional Investor*, 7 November 2016, www.institutionalinvestor.com/article/b14z9p2nzrs2gl/dan-ochs-african-nightmare.

27 Oz Africa Plea Agreement.

28 'Och-Ziff Capital Management admits to role in Africa bribery conspiracies and agrees to pay $213 million criminal fine', Department of Justice, 29 September 2016, www.justice.gov/opa/pr/och-ziff-capital-management-admits-role-africa-bribery-conspiracies-and-agrees-pay-213.

29 Wild, F., Riseborough, J., 'Glencore reviewing bribery allegations said to involve Gertler', *Bloomberg News*, 30 September 2016, www.bloombergquint.com/markets/glencore-reviewing-bribery-allegations-said-to-involve-gertler.

30 'Settlement of dispute with Ventora and Africa horizons', Glencore, 15 June 2018, www.glencore.com/media-and-insights/news/Settlement-of-dispute-with-Ventora-and-Africa-horizons.

31 'Blog: Glencore unfazed by muddy Congo deals', *Global Witness*, 21 May 2014, www.globalwitness.org/en/archive/8622/.

32 'A new mining code for the DRC', *DLA Piper*, 10 August 2018, www.dlapiper.com/en/morocco/insights/publications/2018/08/democratic-republic-of-congo-mining-code/.

Chapter 9 Blood Cobalt

1 Van Reybrouck, D., *Congo: The Epic History of a People* (New York, Ecco Press, 2015), p. 119.

2 Kavanagh, M., 'This is our land', *New York Times*, 26 January 2019.

3 Liwanga, R.C., *Child Mining in An Era of High-Technology, Understanding the Roots, Conditions, and Effects of Labor Exploitation in the Democratic Republic of Congo* (Dearborn MI, Alpha Academic Press, 2017), p. vii.

4 See other studies: Faber et al. estimated that about 23 percent of children worked in the cobalt mining sector, while Bundesanstalt für

Geowissenschaften und Rohstoffe (BGR) found children at 17 mines (or 29 percent); Faber, B., Krause, B., Sánchez de la Sierra, R., 'Artisanal mining, livelihoods, and child labor in the cobalt supply chain of the Democratic Republic of Congo', UC Berkeley CEGA White Papers, 6 May 2017; 'Mapping of the artisanal copper-cobalt mining sector in the provinces of Haut-Katanga and Lualaba in the Democratic Republic of the Congo', BGR, October 2019.

5   Dauvergne, P., *The Shadows of Consumption: Consequences for the Global Environment* (Cambridge MA, MIT Press, 2010), p. 209.

6   'Building a responsible supply chain', *Faraday Insights*, 7 (May 2020), https://faraday.ac.uk/wp-content/uploads/2020/05/Insight-cobalt-supply-chain1.pdf.

7   Banza Lubaba Nkulu, C., Casas, L., Haufroid, V. et al., 'Sustainability of artisanal mining of cobalt in DR Congo', *Nature Sustainability*, 1 (2018), 495–504, https://doi.org/10.1038/s41893-018-0139-4.

8   'Metal mining and birth defects: a case-control study in Lubumbashi, Democratic Republic of the Congo', *The Lancet Planetary Health*, April 2020, www.thelancet.com/journals/lanplh/article/PIIS2542-5196(20)30059-0/fulltext.

9   Van Reybrouck, *Congo*, p. 526.

10   Ibid., p. 527.

11   Carroll, R., 'Return of mining brings hope of peace and prosperity to ravaged Congo', *Guardian*, 5 July 2006.

12   US diplomatic cable, 29 April 2005, *Wikileaks*, https://wikileaks.org/plusd/cables/05KINSHASA731_a.html.

13   Kabemba, C., Bokondu, G., Cihunda, J., 'Overexploitation and injustice against artisanal miners in the Congolese cobalt supply chain', Southern Africa Resource Watch, *Resource Insight*, 18 (January 2020), www.sarwatch.co.za/wp-content/uploads/2020/03/Cobalt-Report-v2-English_compressed.pdf.

14   Quoted in Liwanga, *Child Mining*, p. 22.

15   Clark, S., Smith, M., Wild, F., 'China lets child miners die digging in Congo mines for copper', *Bloomberg News*, 23 July 2008.

16   'This is what we die for, human rights abuses in the Democratic Republic of the Congo power the global trade in cobalt', *Amnesty International*, 19 January 2016, www.amnesty.org/en/documents/afr62/3183/2016/en/.

17   'Exposed: child labour behind smart phone and electric car batteries', Amnesty, 19 January 2016.

18   Pakenham, T., *The Scramble for Africa* (London, Weidenfeld and Nicolson, 1991), p. 588.

19   Jasanoff, M., *The Dawn Watch: Joseph Conrad in a Global World* (London, William Collins, 2017), p. 210.

20   Ibid., p. 213.

21   Ibid., p. 213.

22   Sovacool, B.K., Hook, A., Martiskainen, M., Brock, A., Turnheim, B., 'The decarbonisation divide: Contextualizing landscapes of low-carbon exploitation and toxicity in Africa', *Global Environmental Change*, 60 (2020), 102028.

23   Xing, X., 'A video allegedly staged by UK journalists in DR Congo has nobody fooled about Sino-Congolese relations', *Global Times*, 20 December 2017, www.globaltimes.cn/content/1081258.shtml.

24   Sanderson, H., 'Glencore warns on child labour in Congo's cobalt mining', *Financial Times*, 16 April 2018.

25   'Eurasian Resources Group joins with leading businesses and international organisations to launch the Global Battery Alliance', Eurasian Resources Group, 20 September 2019, https://eurasianresources.lu/en/news/eurasian-resources-group-joins-with-leading-businesses-and-.

26   Sanderson, H., 'NGOs hit out at LME's cobalt sourcing plans', *Financial Times*, 7 February 2019.

27   'Glencore fails to disclose royalty payments for US-sanctioned businessman Dan Gertler', *Resource Matters*, 24 April 2019, https://resourcematters.org/glencore-fails-disclose-royalty-payments-us-sanctioned-businessman-dan-gertler/.

Chapter 10   Dirty Nickel

1   Sun, Y., Burton, M., '"Please mine more nickel," Musk urges as Tesla boosts production', *Reuters*, 23 July 2020.

2   'Chinese owned Ramu NICO brushes aside Basamuk report', Papua New Guinea Today, 12 October 2019, https://news.pngfacts.com/2019/10/chinese-owned-ramu-nico-brushes-aside.html.

3   Doherty, B., 'Rio Tinto accused of violating human rights in Bougainville for not cleaning up Panguna mine', *Guardian*, 31 March 2020.

4   Mudd, G.M., Roche, C., Northey, S.A., Jowitt, S.M., Gamato, G., 'Mining in Papua New Guinea: a complex story of trends, impacts and governance', *Science of The Total Environment*, 741 (2020), 140375.

5   Allen, M., 'A brutal war and rivers poisoned with every rainfall: how one mine destroyed an island', *The Conversation*, 30 September 2020, https://theconversation.com/a-brutal-war-and-rivers-poisoned-with-every-rainfall-how-one-mine-destroyed-an-island-147092.

6   'CEPA: only 80,000 litres slurry escape into Basamuk Sea', *NBC News PNG*, 17 October 2019, www.facebook.com/NBCNewsPNG/posts/952825368404384.

7   Email from Storebrand Asset Management, quoted in Moore, E., 'Why did a Norwegian firm ditch a Chinese company over what it's doing in Papua New Guinea?', *Earthworks*, 12 May 2020, https://earthworks.org/blog/

why-did-a-norwegian-firm-ditch-a-chinese-company-over-what-its-doing-in-papua-new-guinea/.

8   Morse, I., 'Locals stage latest fight against PNG mine dumping waste into sea', *Mongabay*, 22 May 2020, https://news.mongabay.com/2020/05/locals-stage-latest-fight-against-png-mine-dumping-waste-into-sea/.

9   Tesla Q2 2020 earnings conference call transcript, 22 July 2020, Motley Fool, www.fool.com/earnings/call-transcripts/2020/07/23/tesla-tsla-q2-2020-earnings-call-transcript.aspx.

10   Tesla Battery Day presentation, YouTube, 22 September 2020, www.youtube.com/watch?v=l6T9xIeZTds.

11   Huang, C., 'Metallurgical knowledge transfer from Asia to Europe', *Artefact*, 8 (2018), 89–100, http://journals.openedition.org/artefact/1996.

12   Ibid.

13   Christensen, A., 'Thomas Edison – failed geophysicist and prospector?', *LinkedIn*, 1 June 2020, www.linkedin.com/pulse/thomas-edison-geophysicist-prospector-asbj%C3%B8rn-n%C3%B8rlund-christensen?articleId=6672117454019416064.

14   Romney, M., 'America is awakening to China. This is a clarion call to seize the moment', *Washington Post*, 23 April 2020.

15   'EU launches WTO challenge against Indonesian restrictions on raw materials', European Commission, 22 November 2019, https://trade.ec.europa.eu/doclib/press/index.cfm?id=2086&title=EU-launches-WTO-challenge-against-Indonesian-restrictions-on-raw-materials.

16   Bland, B., *Man of Contradictions, Joko Widodo and the Struggle to Remake Indonesia* (Penguin Books, 2020), p. 71.

17   'President Joko met cordially with a delegation led by chairman Chen Xuehua', *Huayou Cobalt*, 9 July 2019, http://en.huayou.com/news/425.html.

18   Ibid.

19   '亲戚合伙撑起一片天:温商项光达、张积敏家族抱团创业的成功样板', 一波说传承有道, 20 June 2018, https://cj.sina.com.cn/articles/view/6034141786/167a9b25a001008ruf.

20   Speech at the China Ningde Stainless Steel New Material Innovation Seminar held in Ningde, Fujian, 21 August 2020. See https://finance.sina.com.cn/stock/stockzmt/2020-08-23/doc-iivhuipp0237616.shtml.

21   'Tsingshan's Indonesia Morowali Industrial Park: build, and they will come', HSBC, reposted from *Caijing Magazine*, 30 (2019), www.business.hsbc.com.cn/en-gb/belt-and-road/story-5.

22   Nickel Mines presentation, May 2021, https://nickelmines.com.au/wp-content/uploads/2021/05/pjn10794-1.pdf.

23   Camba, A., 'Indonesia Morowali Industrial Park: how industrial policy reshapes Chinese investment and corporate alliances', *Panda Paw Dragon*

Claw, 17 January 2021, https://pandapawdragonclaw.blog/2021/01/17/indonesia-morowali-industrial-park-how-industrial-policy-reshapes-chinese-investment-and-corporate-alliances/.

24   See Hudayana, B., Suharko, Widyanta, A.B., 'Communal violence as a strategy for negotiation: community responses to nickel mining industry in central Sulawesi, Indonesia', *Extractive Industries and Society*, 7 (2020), 1547–56.

25   Supriatna, J., 'Deforestation on Indonesian island of Sulawesi destroys habitat of endemic primates', *The Conversation*, 23 October 2020.

26   Supriatna, J., Shekelle, M., Fuad, A.H.H. et al., 'Deforestation on the Indonesian island of Sulawesi and the loss of primate habitat', *Global Ecology and Conservation*, 24 (2020), e10205.

27   'Nickel resources strong bargaining chip for Indonesia: Pandjaitan', *Antara News*, 18 June 2021, https://en.antaranews.com/news/176950/nickel-resources-strong-bargaining-chip-for-indonesia-pandjaitan.

Chapter 11   The Green Copper Tycoon

1   Robert Friedland, keynote speech at the Prospectors and Developers Association of Canada (PDAC) conference in Toronto, March 2020, www.youtube.com/watch?v=h-FbTqJW6eg.

2   Gates, B., *How to Avoid a Climate Disaster: The Solutions We Have and the Breakthroughs We Need* (London, Allen Lane, 2021), p. 41.

3   Ibid., p. 79.

4   See Koelsch, J., 'Chinese firms position for an energy transition copper supercycle', *Baker Institute Blog*, 5 April 2021, http://blog.bakerinstitute.org/2021/04/05/chinese-firms-position-for-an-energy-transition-copper-supercycle/: 'For instance, every thousand battery electric vehicles (BEVs) produced can require approximately 83 metric tonnes (MT) of copper (well more than triple conventional vehicles at 23 MT), while wind turbines incorporate 3.6 MT of copper per megawatt (MW) of output, photovoltaic cells 4-to-5 MT per MW, and flywheels for pumped hydropower 0.3-to-4 MT per MW.'

5   Azadi, M., Northey, S.A., Ali, S.H. et al., 'Transparency on greenhouse gas emissions from mining to enable climate change mitigation', *Nature Geoscience*, 13 (2020), 100–4.

6   'CRU/CESCO-WRAPUP 1 – As copper projects rev up, deficit still seen', *Reuters*, 8 April 2010.

7   Lipton, E., Searcey, D., 'How the US lost ground to China in the contest for clean energy', *New York Times*, 21 November 2021.

8   Isaacson, W., *Steve Jobs: The Exclusive Biography* (New York, Simon & Schuster, 2015), p. 37.

9   Brennan, *The Bite in the Apple*, p. 85.

10 Ibid.

11 Isaacson, *Steve Jobs*, p. 39.

12 McNish, *The Big Score*, p. 26.

13 'Controversial investor makes Burma centrepiece of Asian plan', *Inter Press Service News Agency*, 10 December 1996.

14 Larmer, M., 'At the crossroads: mining and political change on the Katangese-Zambian copperbelt', *Oxford Handbooks Online*, July 2016, www.oxfordhandbooks.com/view/10.1093/oxfordhb/9780199935369.001.0001/oxfordhb-9780199935369-e-20?print=pdf.

15 Wells, J., 'Canada's next billionaire', *MacLeans*, 3 June 1996, https://archive.macleans.ca/article/1996/6/3/canadas-next-billio-naire.

16 Ivanhoe Mines, '2018 news', www.ivanhoemines.com/news/2018/strategic-equity-investment-of-c-723-million-in-ivanhoe-mines-by-china-based-citic-metal-has-been-completed/.

17 Ivanhoe Mines press release, 19 September 2018, https://cn.ivanhoemines.com/news/2018/strategic-equity-investment-of-c-723-million-in-ivanhoe-mines-by-china-based-citic-metal-has-been-completed/.

Chapter 12   The Final Frontier: Mining the Deep Sea

1   McVeigh, K., 'David Attenborough calls for ban on "devastating" deep sea mining', *Guardian*, 12 March 2020.

2   Pavid, K., 'Thank the ocean with every breath you take, says Dr Sylvia Earle', Natural History Museum, 28 November 2017, www.nhm.ac.uk/discover/news/2017/november/thank-the-ocean-dr-sylvia-earle.html.

3   Petersen, S., Krätschell, A., Augustin, N., Jamieson, J., Hein, J.R., Hannington, M.D., 'News from the seabed – geological characteristics and resource potential of deep-sea mineral resources', *Marine Policy*, 70 (2016), 175–87, www.sciencedirect.com/science/article/pii/S0308597X16300732?via%3Dihub.

4   The company asserted that its claims in the Clarion-Clipperton Zone contained enough nickel, copper, cobalt and manganese to potentially build over 250 million electric vehicle batteries. The company later changed its name to The Metals Company, before listing on the stock exchange via reverse merger with Sustainable Opportunities Acquisitions Corp, a special purpose acquisition vehicle.

5   See Mining Watch Canada, Deep Sea Mining Campaign, London Mining Network, 'Why the rush? Seabed mining in the Pacific Ocean', July 2019, www.deepseaminingoutofourdepth.org/wp-content/uploads/Why-the-Rush.pdf, p. 26; and Stutt, A., 'Nautilus Minerals officially sinks, shares still trading', *Mining.com*, 26 November 2019, www.mining.com/nautilus-minerals-officially-sinks-shares-still-trading/.

6   Hein, J.R., Koschinsky, A., Kuhn, T., 'Deep-ocean polymetallic nodules as a resource for critical materials', *Nature Reviews: Earth and Environment*, 1 (2020), 158–69.

7   Rogers, A., *The Hidden Wonders of Our Oceans and How We Can Protect Them* (London, Wildfire, 2019), p. 9.

8   Scales, H., *The Brilliant Abyss: True Tales of Exploring the Deep Sea, Discovering Hidden Life and Selling the Seabed* (London, Bloomsbury Sigma, 2021), p. 96.

9   Davis, J., 'New species from the abyssal ocean hint at incredible deep sea diversity', Natural History Museum, 21 April 2020, www.nhm.ac.uk/ discover/news/2020/april/new-species-from-the-abyssal-ocean-deep-sea-diversity.html.

10   Purser, A., Marcon, Y., Hoving, H-J.T. et al., 'Association of deep-sea incirrate octopods with manganese crusts and nodule fields in the Pacific Ocean', *Current Biology*, 26, 24 (2016), R1268–9.

11   Vonnahme, T.R., Molari, M., Janssen, F. et al., 'Effects of a deep-sea mining experiment on seafloor microbial communities and functions after 26 years', *Science Advances*, 6, 18 (2020).

12   'The Metals Company releases study comparing impacts of land ores to polymetallic nodules', The Metals Company press release, April 2020, https:// metals.co/deepgreen-releases-study-comparing-land-ores-to-nodules/.

13   Hein et al., 'Deep-ocean polymetallic nodules'.

14   'Treasures of the abyss', Geological Society, May 2013, www.geolsoc.org.uk/ Geoscientist/Archive/May-2013/Treasures-from-the-abyss.

15   Mero, J.L., *The Mineral Resources of The Sea* (Amsterdam, Elsevier, 1965), p. 5.

16   Thulin, L., 'During the Cold War, the CIA secretly plucked a Soviet submarine from the ocean floor using a giant claw', *Smithsonian Magazine*, 10 May 2019, www.smithsonianmag.com/history/during-cold-war-ci-secretly-plucked-soviet-submarine-ocean-floor-using-giant-claw-180972154/.

17   CIA Twitter account, 13 May 2019, https://twitter.com/cia/status/ 1128052066650337280?lang=en.

18   Sparenberg, O., 'A historical perspective on deep-sea mining for manganese nodules, 1965–2019', *Extractive Industries and Society*, 6 (2019), 842–54.

19   Arvid Pardo, speech to the United Nations General Assembly, www.un.org/ depts/los/convention_agreements/texts/pardo_ga1967.pdf.

20   ISA website, www.isa.org.jm/about-isa.

Chapter 13   Reduce, Re-use, Recycle: A Closed Loop

1   Skrabec, *The Green Vision of Henry Ford*, p. 175.

2   McGee, P., Sanderson H., 'Electric vehicles: recycled batteries and the search for a circular economy', *Financial Times*, 2 August 2021.

3   'Sustainable Supply Chain for Batteries', Storage X Symposium, Stanford University [online], 2 November 2020, available at https://www.youtube.com/watch?v=FQ0yFAGELnE (accessed 7 April 2022).

4   McGee, Sanderson, 'Electric vehicles: recycled batteries and the search for a circular economy'.

5   Ibid.

6   McDonough, W., Braungart, M., *Cradle to Cradle: Remaking the Way We Make Things* (London, Jonathan Cape, 2008), p. 24.

7   Ibid., p. 25.

8   Ibid., p. 158.

9   Home, A., 'Humble aluminium can shows a circular economy won't be easy', *Reuters*, 26 March 2021.

10 Ibid.

11 'Chemistry can help make plastics sustainable – but it isn't the whole solution', *Nature*, 17 February 2021, www.nature.com/articles/d41586-021-00391-7?utm_source=Nature+Briefing&utm_campaign=4f69ad29e5-briefing-dy-20210219&utm_medium=email&utm_term=0_c9dfd39373-4f69ad29e5-43565849.

12 Mulvaney, D., Richards, R.M., Bazilian, M.D. et al., 'Progress towards a circular economy in materials to decarbonize electricity and mobility', *Renewable and Sustainable Energy Reviews*, 137 (2021), 110604, https://doi.org/10.1016/j.rser.2020.110604.

13 Owen, D., 'The efficiency dilemma', *New Yorker*, 12 December 2010.

14 Smith, B., 'Government won't meet net-zero emissions without "massive change", warns Defra chief scientist', *Civil Service World*, 29 August 2019, www.civilserviceworld.com/professions/article/government-wont-meet-netzero-emissions-without-massive-change-warns-defra-chief-scientist.

15 Owen, 'The efficiency dilemma'.

16 Smil, V., *Growth, From Microorganisms to Megacities* (Cambridge MA, MIT Press, 2019), p. 201.

17 Schmitt, A., 'What happened to pickup trucks?', *Bloomberg News*, 11 March 2021.

18 Platform for Accelerating the Circular Economy, 'A new circular vision for electronics: time for a global reboot', January 2019, https://www3.weforum.org/docs/WEF_A_New_Circular_Vision_for_Electronics.pdf.

19 Statistic courtesy of Analog Devices.

20 McGrath, M., 'Climate change: "dangerous and dirty" used cars sold to Africa', *BBC Online*, 26 October 2020.

21 Xu, C., Dai, Q., Gaines, L. et al., 'Future material demand for automotive lithium-based batteries', *Communications Materials*, 1 (2020), 99, https://doi.org/10.1038/s43246-020-00095-x.

Chapter 14   The World's Greenest Battery

1   Interview for Northvolt Chronicles, available at https://chronicles.northvolt.com/.

2   Campbell, P., 'Eight carmakers on course to miss European CO_2 targets', *Financial Times*, 27 November 2016.

3   Rathi, A., 'Europe is ready to spend billions on batteries to catch up with China', *Quartz*, 15 October 2008.

4   Scott, M., Posaner, J., 'Europe's big battery bet', *Politico*, 26 July 2020, www.politico.eu/article/europe-battery-electric-tesla-china/.

5   Speech by Maroš Šefčovič, European Commission, 23 February 2018, https://ec.europa.eu/commission/presscorner/detail/en/SPEECH_18_1168.

6   Cole, L., 'Breaking new ground: the EU's push for raw materials sovereignty', *Euractiv*, 18 November 2019, www.euractiv.com/section/circular-economy/news/breaking-new-ground-the-eus-push-for-raw-materials-sovereignty/.

7   'Speech by Vice-President Šefčovič at the European Investment Bank (EIB) Board of Directors' meeting', European Commission, 12 June 2019, https://ec.europa.eu/commission/presscorner/detail/en/SPEECH_19_2973.

8   'TIER and Northvolt start partnership to equip e-scooters with greener batteries', Northvolt website, 24 February 2021, https://northvolt.com/articles/tier-scooters/.

9   Sanderson, H., Milne, R., 'Sweden's Northvolt raises $2.75bn to boost battery output', *Financial Times*, 9 June 2021.

10 '如果未来几年仍然是这个趋势，没有投入就没有产出，我们就很难继续待在第一梯队', '技术创新，产业协同共促新能源汽车行业持续健康发展', 29 September 2020, www.catl.com/news/503.html.

11 'A 20,000-tonne oil spill is contaminating the Arctic – it could take decades to clean up', *The Conversation*, 14 July 2020, https://theconversation.com/a-20-000-tonne-oil-spill-is-contaminating-the-arctic-it-could-take-decades-to-clean-up-141264.

12 'An appeal of Aborigen-Forum network to Elon Musk, the head of the Tesla company', Aborigen Forum, https://indigenous-russia.com/archives/5785.

13 'Nornickel announces comprehensive support Programme for the Taimyr's Indigenous Peoples', Nornickel press release, 25 September 2020, www.nornickel.com/news-and-media/press-releases-and-news/nornickel-announces-comprehensive-support-programme-for-the-taimyr-s-indigenous-peoples/.

14 See Foy, H., 'Oligarch Vladimir Potanin on money, power and Putin', *Financial Times*, 13 April 2018.

15 Butt, H., Okun, S., 'Norilsk Nickel plans to expand nickel production to meet growing EV demand', *Fastmarkets*, 20 November 2018, www.metalbulletin.com/Article/3844953/Norilsk-Nickel-plans-to-expand-nickel-production-to-meet-growing-EV-demand.html.

Chapter 15   Cornwall's Mining Revival

1   'Cornish Lithium to receive significant investment from UK government's Getting Building Fund', Cornish Lithium, https://cornishlithium.com/company-announcements/cornish-lithium-to-receive-significant-investment-from-uk-governments-getting-building-fund/.

2   'Debate between Boris Johnson and Steve Double', 16 September 2021, www.parallelparliament.co.uk/mp/boris-johnson/vs/steve-double.

3   Buckley, J.A., *The Cornish Mining Industry, A Brief History* (Redruth, Tor Mark Press, 1992), p. 3.

4   Tonkin, B., 'Heroic and tragic truth behind Poldark: Cornishmen shaped mining in Britain and pushed boundaries the world over', *Independent*, 13 April 2015.

5   Schwartz, S., 'Creating the cult of "Cousin Jack": Cornish miners in Latin America 1812–1848 and the development of an international mining labour market' (1999), https://projects.exeter.ac.uk/cornishlatin/Creating%20the%20Cult%20of%20Cousin%20Jack.pdf.

6   Pell, R., Grant, A., Deak, D., 'Geothermal lithium: the final frontier of decarbonization', Jade Cove Partners, May 2020.

7   See UK Environment Agency report, 'Wheal Jane, a clear improvement', https://consult.environment-agency.gov.uk/psc/tr3-6ee-uk-remediation-ltd/supporting_documents/1992%20EA%20pollution%20incident%20report.pdf.

Conclusion

1   Phillips, B.B., Bullock, J.M., Osborne, J.L., Gaston, K.J., 'Spatial extent of road pollution: a national analysis', *Science of the Total Environment*, 773 (2021), 145589.

2   Dauvergne, P., *The Shadows of Consumption, Consequences for the Global Environment* (Cambridge MA, MIT Press, 2008), p. 215.

3   Hoekstra, 'Underestimated potential'.

4   Smil, *Growth*, p. 511.

5   Figueres, C., Rivett-Carnac, T., *The Future We Choose* (London, Manilla Press, 2020), p. 113.

Index